THE SHORT STORY OF THE UNIVERSE

First published in Great Britain in 2022 by Laurence King,
an imprint of The Orion Publishing Group Ltd,
Carmelite House, 50 Victoria Embankment, London EC4Y 0DZ

An Hachette UK Company

10 9 8 7 6 5 4 3 2 1

A CIP catalogue record for this book is available from the British Library.

ISBN 978-0-85782-938-2

Designed by John Round Design
Origination by DL Imaging, London
Printed in China by C&C Offset Printing Co., Ltd.

p.10 US astronomer Edwin Hubble using the 100-inch Hooker telescope at Mount
Wilson Observatory, Pasadena, California, USA.
p.32 The Hubble Telescope in 2002 when spacecrews installed solar arrays, a
camera, a power control unit, a reaction wheel assembly and an experimental
cooling system. The telescope was first launched and deployed in 1990.
p.60 Comet McNaught photographed over the Australian Outback, 2007.
p.192 Albert Einstein giving a lecture in Vienna in 1921.

Laurence King Publishing is committed to ethical and sustainable production.
We are proud participants in The Book Chain Project
bookchainproject.com®

www.laurenceking.com
www.orionbooks.co.uk

THE SHORT STORY OF THE UNIVERSE

A Pocket Guide to the History, Structure, Theories & Building Blocks of the Cosmos

Gemma Lavender

Laurence King Publishing

Contents

THEORIES

Introduction

EDWIN HUBBLE: 'EQUIPPED WITH HIS FIVE SENSES, MAN EXPLORES THE UNIVERSE AROUND HIM AND CALLS THE ADVENTURE SCIENCE.'

Ever since our distant ancestors first looked up at the night sky and wondered what all those twinkling lights were, we have been mesmerized by the heavens. The stars featured heavily in ancient beliefs: to the Mayans, the Milky Way was the World Tree that linked the underworld with heaven; the Aborigines saw the shape of an emu in the stars; the mythologies of Ancient Greece and Rome have been immortalized in the names of planets, stars and constellations.

Ancient people incorporated celestial bodies into their cultural stories to make sense of the world in the millennia before science overcame astrology. But the urge to understand the Universe is the same today as it always was. We are fortunate to live in an age of science, when we can learn things about those twinkling lights that the ancients could scarcely dream of.

This book shows what science has taught us about the Universe, a vast expanse that is 13.8 billion years old and stretches for at least 93 billion light years and probably much further. It tells the story of our Universe from its distant beginnings to the farthest future, and introduces readers to the laws and forces of the cosmos, as well as to its leading denizens.

Structure

NEIL DEGRASSE TYSON: 'WE ARE ALL CONNECTED: TO EACH OTHER, BIOLOGICALLY; TO THE EARTH, CHEMICALLY; TO THE REST OF THE UNIVERSE, ATOMICALLY.'

When we look out into the Universe, we see structure. Stars are not everywhere; they are bound into vast agglomerations we call galaxies. Those galaxies cluster in groups, and the widest surveys of the cosmos show that these clusters are aligned along enormous filaments made mainly from dark matter. These filaments criss-cross, forming the 'cosmic web' that is the overarching structure of the Universe. On the largest scale, gravity and electromagnetism hold the Universe together. At the other end of the scale, subatomic particles form the foundation of matter. This chapter describes the various lynchpins in cosmic structure and introduces some of the scientists who increased our understanding of it.

History and Future

MARIA MITCHELL: 'DO NOT LOOK AT THE STARS AS BRIGHT SPOTS ONLY. TRY TO TAKE IN THE VASTNESS OF THE UNIVERSE.'

If the history of the Universe were condensed into one year, homo sapiens would have arisen at 11:52 PM on 31 December. We are but a recent footnote in the story of cosmic evolution.

The history of our Universe was dictated by events that occurred in an even shorter time span – the first second after the Big Bang. It is in this very early era, when the disciplines of cosmology, particle physics and quantum physics collided, that the family tree of particles grew, and the fundamental forces governing them came into their own. The Universe was still small enough that quantum effects could affect the entire cosmos, with repercussions that endure to this day.

Components

MARTIN REES: 'THERE ARE AT LEAST AS MANY GALAXIES IN OUR OBSERVABLE UNIVERSE AS THERE ARE STARS IN OUR GALAXY.'

When scientists take an inventory of the Universe, they must include all its mass and energy. Some 69 per cent of the Universe is made from dark energy, a mysterious force that causes the expansion of the Universe to accelerate. Another 26 per cent of the Universe is made from dark matter, a substance that does not interact with light. Finally, a mere 5 per cent of the Universe is ordinary matter that we can see, smell and touch. This 5 per cent makes up the trillions of galaxies, stars, planets, moons, asteroids and interstellar dust and gas that we can observe stretching all the way to the cosmic horizon. This chapter describes these and other types of object in the Universe that make up the 5 per cent.

Theories

CARL SAGAN: 'IF YOU WANT TO MAKE AN APPLE PIE FROM SCRATCH, YOU MUST FIRST CREATE THE UNIVERSE.'

Here we look at how the structure, history and components of the Universe tie together. The word 'theory' is often misunderstood. A theory isn't just an idea – it's a conceptual framework that supports predictions based on observations and a body of evidence that supports that theory. From the detective stories behind our understanding of cosmic inflation, stellar evolution and planet formation to the inspiration of Einstein's theories of light and gravity, it's all here in this final section.

Distance units

The scale of the Universe inevitably involves huge distances, and objects at scales vastly larger than those we encounter in everyday life. Throughout this book, therefore, we use some of the following units for brevity and readability.

Astronomical Unit (AU): Used in measurements of solar systems, this is roughly the mean distance at which Earth orbits the Sun – 149.6 million kilometres (93 million miles).

Light year: This widely used unit is equivalent to the distance travelled by light in one Earth year – roughly 9.5 million million kilometres (5.9 trillion miles).

Time

The history of the Universe takes place across billions of years, but the timescales can vary widely. Some crucial events took place just fractions of a second after the Big Bang; other processes took millions or even billions of years. In the History chapter, we date events based on the time elapsed since the Big Bang. Since time began with the Big Bang, we refer to the Big Bang as being at T (time) = 0. Every subsequent event is T +.

Scientific notation

When dealing with very large or small numbers, we occasionally use the form $a \times 10^b$. In this scientific notation (used to reduce the number of digits needed to write very large or small numbers), a is the 'mantissa' (the most significant figures in the number), while b is the 'exponent' – the number of times 10 would need to be multiplied by itself in order to represent the decimal equivalent.

For example: $10^3 = 10 \times 10 \times 10 = 1,000$, so 3,400 can be written as 3.4×10^3.

Note that a – sign before the exponent represents a negative power thus:

$10^{-b} = 1/10b$

So for example $10^{-6} = 1/1,000,000$, and 0.000003 can be written as 3×10^{-6}.

How to use this book

This book is divided into four parts – Structure, History and Future, Components, and Theories – each exploring a different way of looking at the Universe: what it constitutes; the timeline of cosmic history; the bodies that inhabit it; and the laws and forces that govern it. There's information on astronomers, details of important moments in astronomical history, and useful cross-referencing.

Key date
Key scientists
Key developments
Cross-references to Components and Theories

Scientist background information

Notable examples

Cross-references to Structure, History and Future and Theories

Structure

The Universe

EDWIN HUBBLE: *A SPIRAL NEBULA AS A STELLAR SYSTEM, MESSIER 31*
MOUNT WILSON OBSERVATORY, USA • 1924–29

Key publications

Edwin Hubble, *Extragalactic Nebulae*, 1926

Edwin Hubble, *A Spiral Nebula as a Stellar System, Messier 31*, 1929

Edwin Hubble, *A Relation Between Distance and Radial Velocity Among Extragalactic Nebulae*, 1929

The Universe is everything around us; it is the entirety of existence. It is filled with matter that congregates to form asteroids, moons, planets, stars, galaxies, and clusters of galaxies, all surrounded by mysterious dark matter.

The Universe is old. It began 13.8 billion years ago in the Big Bang, and has been expanding ever since. In 1998 astronomers discovered that the rate of this expansion is accelerating, driven by some unknown force that we don't understand, and which scientists have labelled 'dark energy'.

And the Universe is huge. For the parts of the cosmos that we can see – known as the Observable Universe – estimates suggest that space extends for 93 billion light years across. The horizon, or edge, of this Observable Universe is defined as the distance for which light has had time to reach us during the history of the Universe. There's still much more of the Universe over that cosmic horizon, but the Universe is so big that light hasn't yet had time to reach us from there.

To get a feel for the size, one light year is approximately 9 trillion kilometres (6 trillion miles). Compare this to the average distance between Earth and the Sun, which is 149.6 million kilometres (91.4 million miles), or the distance of the outermost planet, Neptune, from the Sun, which is 4.5 billion kilometres (2.8 billion miles).

EDWIN HUBBLE, 1889–1953

Hubble was an accomplished athlete, studied law, worked as a schoolteacher and joined the army before becoming an astronomer. Working at Mount Wilson Observatory in California, he transformed our understanding of the Universe, discovering that there were galaxies beyond our own, and that the Universe is expanding. The Hubble Space Telescope is named after him.

SPIRAL GALAXIES **p.66** THE BIG BANG THEORY **p.194**

Albert Einstein (far left) visiting Mount Wilson Observatory,
California, USA, c.1931. Edwin Hubble can be seen right at the
back, second from left.

COSMIC INFLATION **p.195** DARK ENERGY **p.215**

Spacetime

ALBERT EINSTEIN: *THE FIELD EQUATIONS OF GRAVITATION* • BERLIN, GERMANY, 1915

The Universe around us has four dimensions – three spatial dimensions (length, breadth and width) and one dimension of time. Albert Einstein called this 'spacetime', and his Theory of General Relativity showed how spacetime behaves under the influence of mass and gravity.

To make sense of spacetime, Einstein imagined it as a huge sheet of rubber, bending and wrinkling when hefty planets, stars and galaxies are placed on it. The more mass an object has, the greater its gravity and the more it tugs and warps the space around it.

'Matter tells space how to curve, and curved space tells matter how to move,' the physicist John Wheeler (1911–2008) eloquently stated. This has consequences for light, which travels through space, following its curvature. Where spacetime bends, the path of light also bends. This leads to the phenomenon of gravitational lenses, where the curvature of spacetime caused by the gravity of a massive object magnifies the light of a more distant object.

Einstein also realized that, just as gravity affects space, it can also affect time. In the strongest gravitational fields, clocks run more slowly. If you fell into a black hole, you wouldn't notice your wristwatch ticking more slowly, but an observer outside the black hole, in a lesser gravitational environment, would see things happening to you in slow motion.

Key publications
Albert Einstein, *The Field Equations of Gravitation,* 1915
Hermann Minkowski, *Space and Time,* 1908–09
Orest Khvolson, *Über eine mögliche Form fiktiver Doppelsterne,* 1924

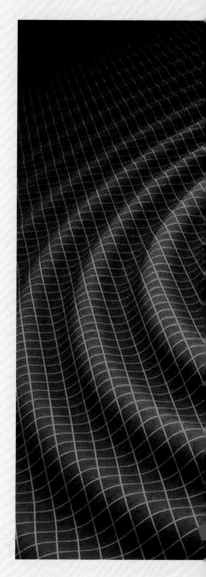

An artist's impression of gravitational waves emitted by an inwardly spiralling pair of neutron stars that are about to collide.

STELLAR BLACK HOLES **p.122** SPECIAL RELATIVITY **p.196**

ALBERT EINSTEIN, 1879–1955
Possibly the greatest scientist, and certainly the most famous, of all time, Albert Einstein is a household name. His theories of relativity (both Special and General) taught us about light and energy, space and gravitation, and then applied them to the Universe. Ironically, he won the Nobel Prize in 1921 for something else – discovering the photoelectric effect.

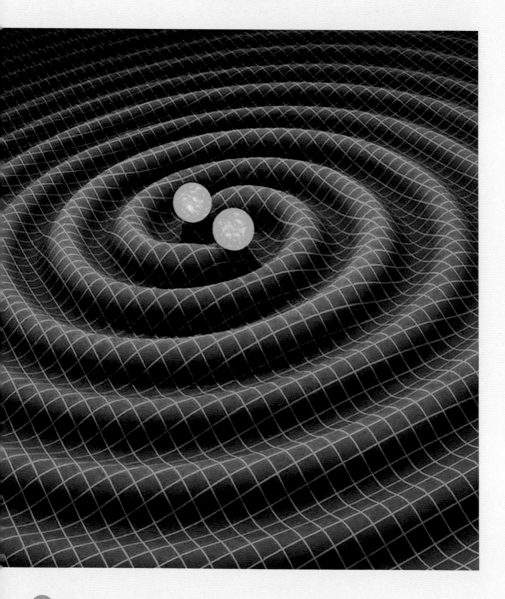

GENERAL RELATIVITY **p.197**

Distribution of Matter

THE VIRGO CONSORTIUM: *SIMULATIONS OF THE FORMATION, EVOLUTION AND CLUSTERING OF GALAXIES AND QUASARS* • GERMANY AND THE UK • 2005

R. BRENT TULLY, 1943–
Richard Brent Tully's name will live on in the annals of astrophysics history in the form of the Tully–Fisher Relation, which he developed with the astronomer James Richard Fisher (b.1943), and which describes a correlation between how luminous a spiral galaxy is, and how fast it is rotating. Tully also discovered the Pisces–Cetus Supercluster Complex, which includes our galaxy and is a giant wall of galaxies that forms part of the cosmic web.

There's a lot of empty space in space. Looking out into the night sky is enough to tell you that – there are big gaps between the stars, and if we look further out, big gaps between the galaxies. On the largest scales, there are huge voids, millions of light years across, that contain very few galaxies, while elsewhere there are regions where galaxies are clustered together in their hundreds or thousands.

The way in which matter is distributed throughout the Universe tells us much about how the Universe has evolved, and the role that gravity has had to play in that. The Big Bang was not smooth: when it created all the matter and energy in the Universe, it was slightly lumpy. The denser, lumpier regions had slightly more gravity than other, less lumpy regions, and so they attracted more and more matter to them. Soon, most of the matter in the Universe came to be in these denser regions, which form a 'cosmic web' of giant filaments of matter and dark matter that criss-cross the Universe. Galaxies form along these filaments, and where the filaments meet we find galaxy clusters. Many galaxy clusters together can form walls, or sheets of galaxies, which are the largest structures in the Universe. The distribution of matter in the Universe today, thanks to gravity, can tell us much about the conditions in the Big Bang that ultimately led to this distribution.

Key publications
Margaret Geller and John Huchra, *Mapping the Universe,* 1989
R. Brent Tully, Hélène Courtois, Yehuda Hoffman & Daniel Pomarède, *The Laniakea Supercluster of Galaxies,* 2014
Volker Springel *et al, Simulating the Joint Evolution of Quasars, Galaxies and their Large-Scale Distribution,* 2005

LIGHT AND MATTER SEPARATE **p.41** FILAMENTS AND VOIDS **p.62**

The matter content of the Universe.
Matter makes up 30 per cent of the
Universe, the rest being dark energy.
The matter can be dark (29 per cent)
or visible (1 per cent, top layer). Dark
matter can be further divided into non-
baryonic (25 per cent, bottom layer) and
baryonic (4 per cent, middle layer).

CLUSTERS AND SUPERCLUSTERS **p.64** GALAXY EVOLUTION **p.212** DARK MATTER **p.214**

Bound Systems

FRITZ ZWICKY: *ON THE MASSES OF NEBULAE AND OF CLUSTERS OF NEBULAE* • CALTECH, USA • 1933

Key publications

Vera Rubin and Kent Ford, *Rotation of the Andromeda Nebula from a Spectroscopic Survey of Emission Regions,* 1970

Fritz Zwicky, *On the Masses of Nebulae and of Clusters of Nebulae,* 1937

Our Solar System of planets is a gravitationally bound system, the eight planets and multitude of comets and asteroids all in thrall to the Sun's gravity. In turn, the Sun is bound by gravity to our Milky Way galaxy.

On larger scales, galaxies are bound to one another gravitationally to form galaxy clusters. The Milky Way is in a little cluster called the Local Group, along with the Andromeda galaxy, the Triangulum galaxy and a smattering of smaller, dwarf galaxies. The Local Group is gravitationally bound to the larger Virgo–Coma cluster of galaxies, which in turn is gravitationally bound to other clusters to form the Pisces–Cetus Supercluster Complex. Even larger than that is the Laniakea Supercluster, which gravitationally holds several smaller superclusters.

The motion of planets, stars or galaxies in bound systems tells us about how much gravity they feel, which in turn is related to how much mass there is. In 1933, Fritz Zwicky (1898–1974) noticed that galaxies in galaxy clusters were moving faster than they should be for the amount of visible mass. In the 1960s, Vera Rubin found that stars and nebulae in galaxies were doing the same thing. They should have been flying away instead. The conclusion was that there must be some invisible mass – dark matter – providing additional gravity.

VERA RUBIN, 1928–2016

Becoming an astronomer in an era when women doing astronomy was frowned upon, Vera Rubin helped to pave the way for many female astronomers who followed in her footsteps. Her PhD dissertation described how galaxies clustered together in bound systems, but her male-led colleagues did not follow up on this idea until the 1970s. It was during that time that Rubin and Kent Ford (b.1931) discovered evidence for dark matter.

MERGING GALAXIES **p.46** UNIVERSAL GRAVITATION **p.201**

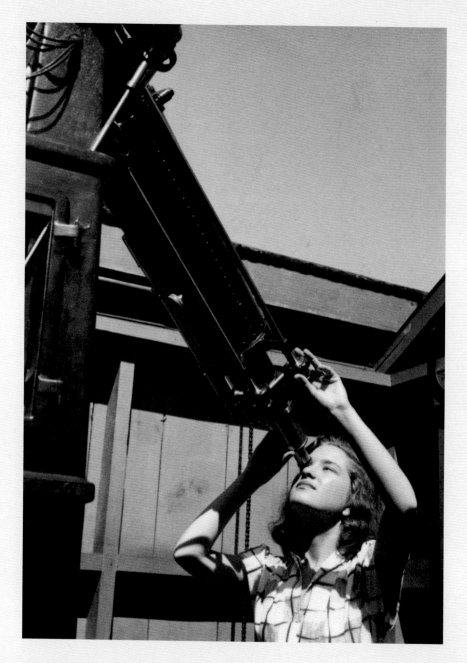

Vera Rubin at Vassar College, New York State, from which she
was the only student in her year to graduate in astronomy.

DARK MATTER **p.214**

Diffuse Matter

VESTO SLIPHER: *PECULIAR STAR SPECTRA SUGGESTIVE OF SELECTIVE ABSORPTION OF LIGHT IN SPACE* • LOWELL OBSERVATORY, USA • 1909

COSMIC RECYCLING **p.48**

Williamina Fleming (foreground, seated) with other members of Edward Pickering's research team, a group known as the Harvard Computers. Other members included Henrietta Swan Leavitt, Annie Jump Cannon and Antonia Maury.

Key publications
Vesto Slipher, *Peculiar Star Spectra Suggestive of Selective Absorption of Light in Space,* 1909
Vesto Slipher, *On the Spectrum of the Nebula in the Pleiades,* 1912

Between the stars, and between the galaxies, space is not as empty as it appears. It's filled with gas, both hot and cold, and lots of dust produced by stellar evolution. In fact, roughly 15 per cent of our galaxy's visible constituents are taken up by interstellar gas and dust, a notable chunk of matter in the Milky Way.

The interstellar medium – that is, the stuff between the stars – is comprised of roughly 99 per cent gas and a smattering of dust. Three-quarters of that gas is hydrogen, while the remainder is mostly helium, with a smattering of other elements. On average, there is a single atom or molecule of gas, mostly hydrogen, for every cubic centimetre of interstellar space. In comparison, air on Earth offers 30 quintillion (that's 30,000,000,000,000,000,000) molecules in the same volume.

Meanwhile, if interstellar dust is particularly thick, it will block the light of stars behind it. If not, it might be heated up and start glowing, allowing us to see the nebulous birthplace of stars. In other cases, starlight can often be reflected off diffuse clouds of dust – in this case, we observe a reflection nebula.

Gas in the intergalactic medium between the galaxies can be incredibly hot, reaching temperatures of up to 10 million degrees Celsius in galaxy clusters, causing it to radiate primarily in X-rays.

WILLIAMINA FLEMING, 1857–1911
One of the famous Harvard Computers – women astronomers working for Edward Pickering (1846–1919) – Fleming specialized in classifying stars through their spectra, which became the basis for subsequent stellar classification models that help us to understand the nature of stars. She also discovered 59 nebulae, including in 1888 the well-known Horsehead Nebula in Orion, which is filled with interstellar gas and dust.

ORIGINS OF THE SOLAR SYSTEM **p.49** BOK GLOBULES **p.92**

Stars

ANNIE JUMP CANNON: *SPECTRA HAVE BRIGHT LINES* • HARVARD COLLEGE
OBSERVATORY, USA • 1916

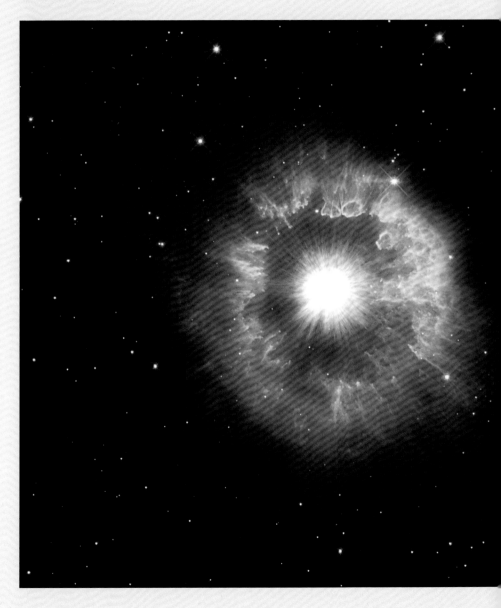

THE FIRST STARS **p.43** THE POWER SOURCE OF STARS **p.206** →

The massive, luminous star AG Carinae, imaged by the Hubble Space Telescope.

In our galaxy alone, it's been calculated that there are around 100 billion stars. Beyond its confines, there are an estimated 1 billion trillion stars visible throughout the observable Universe. Stars are bright, giant balls of mostly hydrogen gas – and our own Sun is one of them.

Of those 1 billion trillion stars, we can see at most about 3,000 in the night sky with the naked eye. The brightest of these make up the constellations, which are patterns of stars as we see them in the night sky. There are 88 constellations, split between the northern and southern celestial hemispheres.

On average, stars in our Milky Way are about 4 light years apart; our nearest stellar neighbour, Proxima Centauri, is 4.24 light years away. It's a red dwarf star, the smallest, coolest and least massive kind of star. Other stars can weigh hundreds of times the mass of the Sun, and be hundreds of times larger, and tens of thousands of times brighter.

Stars are different from other objects like planets because they are able to generate their own energy through powerful nuclear fusion reactions in their hot cores. It's the energy from these reactions that makes stars shine, and makes the Sun breathe life on planet Earth.

Key publications

Antonia Maury and Edward Pickering, *Spectra of bright stars photographed with the 11-inch Draper Telescope as part of the Henry Draper Memorial,* 1897

Ejnar Hertzsprung, *Über die Sterne der Unterabteilungenc undac nach der Spektralklassifikation von Antonia C. Maury,* 1908

Julianna Sackmann, Arnold Boothroyd and Kathleen Kraemer, *Our Sun. III. Present and Future,* 1993

MARGARET BURBIDGE, 1919–2020
A pioneering British astronomer, Margaret Burbidge was one quarter of the famous B^{2}FH paper that described how stars produce the chemical elements by way of nuclear reactions in their core – reactions that provide the energy to power a star. The other members of the quartet were her husband, Geoffrey Burbidge (1925–2010), William Fowler (1911–95) and Fred Hoyle (1915–2001).

STELLAR EVOLUTION **p.207** STELLAR NUCLEOSYNTHESIS **p.208**

Non-luminous Objects

GERARD KUIPER: *SURVEY OF PLANETARY ATMOSPHERES* • MCDONALD OBSERVATORY, TEXAS, USA • 1949

Everything in the Universe can be placed into one of two categories: objects that produce their own light, and objects that don't. The Sun, like every other star, shines from the energy produced by nuclear reactions in its core. The planet Jupiter, on the other hand, doesn't produce its own light, but reflects that of the Sun. That's what we see when we look at planets in the night sky – reflected sunlight.

Planets, moons, asteroids, comets – these are non-luminous objects. How much light is reflected by the surface of a planet or moon is known as its albedo. Tough, rocky celestial bodies often absorb more light than they reflect, leaving them with a low albedo; our Moon – while it appears bright because it is close to us – reflects a mere 7 per cent of the light our Sun throws at it (we say it has an albedo of 0.07). On the other hand, icy bodies such as Jupiter's moon Europa are highly reflective, with albedos as high as 0.6 or more.

Reflected light is useful to scientists. The albedo, and the wavelengths reflected, can tell us about the type of surface doing the reflecting, and its composition. Molecules in the atmospheres of these bodies can also absorb some of the reflected light at specific wavelengths before it reaches us, leaving gaps in its spectrum at those wavelengths that scientists can identify and attribute to those molecules.

SVANTE ARRHENIUS (1859–1927)
Svante Arrhenius, who won the Nobel Prize for Chemistry in 1903, helped to popularize the notion that Venus was a lush swamp land, based on its highly reflective clouds, which he surmised were water vapour. In reality, there is no water on Venus, and the clouds are carbon dioxide. However, Arrhenius's real legacy was to show how carbon dioxide is a greenhouse gas.

BIRTH OF THE MOON **p.52** VENUS **p.132**

Key publications
Gerard Kuiper, *Survey of Planetary Atmospheres*, 1949
Carl Sagan, *Physical Studies of the Planets*, 1960
Stephen H. Dole and Isaac Asimov, *Planets for Man*, 1964
Peter Ward and Donald Brownlee, *Rare Earth*, 2000

Earth's Moon shines by reflected light from the Sun. Here
we see two sides to the Moon – the illuminated side, and the
un-illuminated hemisphere.

MARS **p.144** JUPITER **p.154**

Elements

MARGARET BURBIDGE, GEOFFREY BURBIDGE, WILLIAM FOWLER AND FRED HOYLE: *SYNTHESIS OF THE ELEMENTS IN STARS* • UNIVERSITY OF CAMBRIDGE, UK, AND CALTECH, USA • 1957

CECILIA PAYNE-GAPOSCHKIN, 1900–79
Being the most common elements in the Universe, hydrogen and helium are also the primary elements of stars, a fact that was discovered in 1925 by Cecilia Payne-Gaposchkin. Other astronomers initially tried to dissuade her from coming to this conclusion as it contradicted the consensus at the time, but it took only four years for Payne-Gaposchkin to be proven correct.

The most abundant element in the entire Universe is hydrogen, accounting for 74 per cent of all matter. It's a simple atom, comprised of two subatomic particles, an electron and a proton, and as such it is the lightest atom in the Periodic Table.

Hydrogen was formed during the first three minutes of the Universe, and it wasn't alone: it was joined by helium and a little bit of lithium at the same time. Helium is the second most common element in the Universe, making up roughly 24 per cent of it.

There are 92 naturally occurring elements in the Periodic Table, and these other elements came along later, thanks to the stars. The nuclear reactions in Sun-like stars create carbon, nitrogen and oxygen, while more massive stars form heavier elements such as silicon and iron. When these massive stars explode, more elements such as nickel and zinc are forged in the fires of the supernovae, while studies of colliding neutron stars – which are the dense leftover cores of stars that have gone supernova – can produce precious metal elements such as gold and platinum. Ultimately, all of the elements are made in space.

Key publications
Cecilia Payne-Gaposchkin, *Stellar Atmospheres; A Contribution to the Observational Study of High Temperature in the Reversing Layers of Stars*, 1925
Ralph Alpher, Hans Bethe and George Gamow, *The Origin of Chemical Elements*, 1948
B²FH, *Synthesis of the Elements in Stars*, 1957
The LIGO Consortium et al, *Multi-Messenger Observations of a Binary Neutron Star Merger*, 2017

THE FIRST THREE MINUTES **p.38** FORGING THE ELEMENTS **p.39**

The Crab Nebula is the remnant of a supernova, rich in heavy
elements formed both inside the star that exploded and in the
violence of its supernova.

COSMIC RECYCLING **p.48** STELLAR NUCLEOSYNTHESIS **p.208**

Subatomic Particles

J. J. THOMSON: *CATHODE RAYS* • CAMBRIDGE, UK • 1897

Key publications

Ernest Rutherford, *Collision of α particles with light atoms. IV. An Anomalous Effect in Nitrogen,* 1919

Peter Higgs, *Broken Symmetries and the Masses of Gauge Bosons,* 1964

Murray Gell-Mann, *A Schematic Model of Baryons and Mesons,* 1964

Zoom into the structure of an atom and you'll find that it's made up of smaller particles: electrons, protons and neutrons. We refer to these as subatomic particles – particles smaller than an atom. They were made shortly after the cosmos sprang into existence almost 13.8 billion years ago.

Scientists have discovered that the Universe is made up of a zoo of particles. They're organized according to the Standard Model, which is divided into two broad flavours: fermions and bosons.

Fermions make up the matter that we see, touch, smell and taste here on Earth, and they are further divided into leptons and quarks. The most familiar lepton is the negatively charged electron. Each kind of lepton also has a partner called a neutrino that rarely interacts with anything since it has no charge and very little mass.

The quarks are the building blocks of protons and neutrons. There are six kinds of quark: up, down, charm, strange, top and bottom. When three quarks are bound together, we call that entity a hadron. One up quark and a couple of down quarks make a neutron, while two up quarks and one down quark comprise a proton.

And then there are the bosons, of which the photon is one. They carry the forces that govern nature and also feature the gluon and the Z and W gauge bosons. Thrown into the mix is the famous Higgs boson, which gives everything in the known Universe its mass.

FROM ENERGY TO MATTER **p.37** →

PETER HIGGS

Born in 1929 in Newcastle, UK, Peter Higgs became famous around the world when his long-predicted particle, the Higgs Boson, was discovered in 2012 by the Large Hadron Collider. However, Higgs had led an award-filled research career before then, in his position at the University of Edinburgh.

A depiction of the structure of an atom, with the nucleus of protons and neutrons in the middle, and electrons orbiting around them.

THE FIRST THREE MINUTES **p.38** LIGHT AND MATTER SEPARATE **p.41**

Fundamental Forces

ISAAC NEWTON: *PHILOSOPHIÆ NATURALIS PRINCIPIA MATHEMATICA*
CAMBRIDGE, UK • 1687

If it wasn't for the fundamental forces, everything in the Universe would fly apart, and no reactions could take place. There are four forces in total: the weak and strong forces, which play a part in interactions on an atomic level, and the wider-reaching forces – gravity and the electromagnetic force. Together, they unify nature.

The forces are carried by particles named bosons. The photon carries the electromagnetic force. Gluons carry the strong force, which holds together particles such as the protons and neutrons that form the nuclei of the atoms in our bodies. The W and Z gauge bosons carry the weak force, which is at the heart of the radioactive decay of atoms. Gravity is believed to be carried by a theoretical particle named a graviton, but scientists have yet to discover gravitons in nature.

Electromagnetism is a force that acts between charged particles, and is transmitted by an electromagnetic wave in the form of a photon (in quantum mechanics, light has the bizarre property of behaving like a wave and a particle).

Although gravity might keep our feet on the ground and the planets in orbit around the Sun, it's also considered to be the weakest force (the strong and weak forces are far stronger, albeit only over subatomic distances).

MICHAEL FARADAY, 1791–1867
One of the greatest ever physicists and chemists, Michael Faraday had no formal scientific training and was self-taught, yet went on to make crucial discoveries about the relationship between electricity and magnetism as an electromagnetic force. He was also passionate about teaching science to the public, beginning the Christmas Lectures at London's Royal Institution that continue to this day.

Key publications
Albert Einstein, *The Field Equations of Gravitation*, 1915
Enrico Fermi, *Versuch einer Theorie der β-Strahlen. I,* 1934

THE BIG BANG **p.35** THE MOMENT OF INFLATION **p.36** →

Michael Faraday played a key role in modern understanding
of electromagnetism.

GENERAL RELATIVITY **p.197** UNIVERSAL GRAVITATION **p.201**

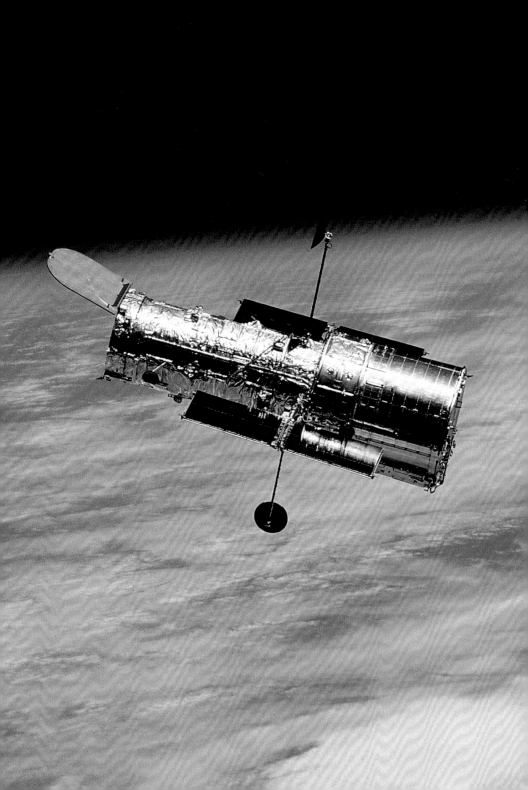

History and Future

Before the Beginning

Before time

KEY SCIENTISTS: ANDREI LINDE • LEE SMOLIN • PAUL STEINHARDT
NEIL TUROK • MARTIN BOJOWALD

An artist's impression of the Oscillating Universe Theory.

Was the Big Bang the birth of time and space, or was there another universe that existed before ours? Many scientists believe that everything began with the Big Bang 13.8 billion years ago. However, some scientists are hooked on the idea of another cosmos before our own, that our Universe was born from a pre-existent one, budding off a sea of rapidly inflating space from which countless other universes also sprang, possibly with different properties from our own. This is the theory of Eternal Inflation.

There are other theories, too. One is the concept of a cyclical universe, detailed in Loop Quantum Cosmology and some forms of String Theory, wherein a previous universe expanded much as ours is doing, before reaching some critical turning point, when it began to contract in a 'Big Crunch'. At the Big Crunch – a kind of reverse Big Bang – all the matter and energy of the

Universe squeezed into a single point. The Universe then rebounded and began inflating again, to begin the cycle once more.

Another hypothesis is the Oscillating Universe Theory, according to which the Universe in which we live now is somewhere between a Big Bang and a Big Crunch.

KEY DEVELOPMENT
In the late 1970s, Alan Guth (b.1947) developed the concept of inflation, which explained how our Universe grew from a single point into a macroscopic cosmos. However, scientists couldn't figure out how inflation stopped, until Andrei Linde (b.1948) came up with the idea of Eternal Inflation in 1984. In this theory, inflation only stops in certain places; where it does, bubble universes form, and our Universe would be one such bubble.

THE BIG BANG **p.35**   THE MOMENT OF INFLATION **p.36**   THE FATE OF THE UNIVERSE **p.59**
MULTIVERSE **p.198**

The Big Bang

KEY SCIENTISTS: EDWIN HUBBLE • GEORGES LEMAÎTRE • RALPH ALPHER
GEORGE GAMOW • ROBERT WILSON • ARNO PENZIAS

T = 0

An artist's impression of the Big Bang, which spawned the Universe.

Our Universe started from a single, small phenomenon known as a singularity – a one-dimensional point that contained all the mass and energy of the Universe in an incredibly tiny space. It's a place where space and time – at least as we understand them today – have no meaning. Even Einstein's Theory of Relativity fails to explain it. What happened next was a sudden event: the Big Bang. But despite its name, it wasn't a loud explosion, it was a violent 'stretch' that pulled spacetime into existence.

There is evidence for the Big Bang, which was initially noted in 1929 when astronomer Edwin Hubble – who was peering through the eyepiece of the 2.5-metre (100-inch) Hooker telescope at Mount Wilson Observatory in California – realized that light waves emanating from faraway galaxies were being stretched into longer, redder wavelengths. This meant that space was being stretched and that the Universe was expanding. This confirmed the suspicions of astronomer and Belgian Catholic priest Georges Lemaître (1894–1996), who was adamant that our Universe must have begun from a single point, one that could be traced back in time.

KEY DEVELOPMENT
The concept of the Big Bang came from Edwin Hubble's discovery that the Universe is expanding, with all the galaxies moving away from each other. If the Universe is expanding and getting bigger, then logically it must have been smaller at some point in the past. Taken to its ultimate conclusion, we can rewind history to show that the Universe must have been just the size of a point 13.8 billion years ago.

THE BIG BANG THEORY **p.194** REDSHIFT AND THE DOPPLER EFFECT **p.203**

The Moment of Inflation

$T = 10^{-35}$ sec

KEY SCIENTISTS: ALAN GUTH • ANDREI LINDE • ALEXEI STAROBINSKY
PAUL STEINHARDT

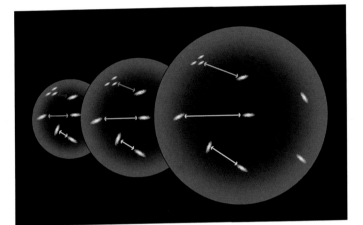

Inflation explains why opposite sides of the Universe look pretty much the same, despite being too far apart for light to communicate between them.

The visible Universe, from one side to the other, stretches about 93 billion light years. Yet the Universe is only 13.8 billion years old, and given that the maximum speed limit is the speed of light, and one light year is the distance light travels in one year, then there hasn't been enough time for light and radiation to reach from one side of the Universe to the other. Yet there's a conundrum, because one side of the Universe looks pretty much like the other side, as though they were in contact in the distant past. The Universe is also very flat – parallel lines remain parallel to infinity, as though something has smoothed out the Universe. Cosmologists call these the horizon problem and the flatness problem, respectively.

In the late 1970s, physicist Alan Guth came up with a solution, called inflation. He described an era, just 10^{-35} seconds after the Big Bang, during which the Universe inflated incredibly quickly, expanding a trillion trillion times, faster than the speed of light. Then, just as quickly as this period of rapid inflation began, it stopped just 10^{-34} seconds after the Big Bang.

KEY DEVELOPMENT

Testing how 'flat' the Universe is has been a key part of cosmological experiments. NASA's Wilkinson Microwave Anisotropy Probe (WMAP) and the European Space Agency's Planck mission have both shown that spacetime is perfectly flat, to an accuracy of 1 per cent. Space is therefore not curved – at least not on the scale of the visible Universe: if you set off in one direction, space will not curve around to return you to your starting point.

COSMIC INFLATION **p.195**

From Energy to Matter

KEY SCIENTISTS: ANDREI LINDE • ALEXEI STAROBINSKY • LEV KOFMAN
ALEXANDER DOLGOV

Possibly the most famous equation in physics is Albert Einstein's $E=mc^2$, which tells us that energy and mass (i.e. belonging to matter) are equivalent. Raw energy is able to create matter.

The Universe began as raw energy. It rapidly inflated for a fraction of a second, and when inflation was over, the Universe was trillions of times bigger in volume. Yet, as we know, if something expands, it cools. The Universe should have been very cold at the end of inflation, but George Gamow (1904–68) showed in the 1940s that the Big Bang should have been a primeval fireball, for it would only be in very hot conditions that Big Bang nucleosynthesis could take place to create the first elements.

The existence of Cosmic Microwave Background radiation also tells us that the Big Bang must have been very hot. But in the 1990s Andrei Linde (b.1948), among others, showed how the energy of inflation was dumped back into the Universe as heat, which 'reheated' the Universe. As the Universe expanded more slowly now, the temperature dropped more gradually. This release of the inflationary field energy would have also converted into the fundamental particles that eventually built up matter as we know it.

KEY DEVELOPMENT
Our modern understanding of inflation revolves around something called Quantum Field Theory (QFT), which describes the behaviour of subatomic particles and the interactions between them. QFT was developed from the 1920s onwards by combining classical field theories (such as electromagnetism) with quantum physics and relativity.

The physicist Andrei Linde, whose breakthroughs have defined the science of the early Universe.

COSMIC INFLATION **p.195**

The First Three Minutes

T = 3 min

KEY SCIENTISTS: STEVEN WEINBERG • GEORGE GAMOW • ROBERT HERMAN
RALPH ALPHER

Although we don't know what caused the Big Bang, it is remarkable that we know in great detail what happened in the brief period that followed. This was detailed in physicist Steven Weinberg's (b.1933) famous 1977 book *The First Three Minutes*.

The initial high temperature after reheating meant that photons could smash protons and neutrons apart, preventing them from joining to create atomic nuclei. But after two minutes had elapsed, the Universe's temperature had dropped to a little over a billion degrees Celsius. This was 'cool' enough that photons couldn't pick up enough speed to smash protons and neutrons apart. Instead, they were the perfect conditions for the making of deuterium nuclei, which form through the union of a proton and neutron and are held together by strong nuclear force.

Three minutes in, and the Universe was continuing to cool, falling below a billion degrees: conditions were just right for protons and neutrons to bind together to make helium nuclei. Many more protons remained unbound, but a single proton comprises a hydrogen nucleus. This set the scene for a Universe dominated by hydrogen and helium.

KEY DEVELOPMENT

Much of the theory behind the first 3 minutes and Big Bang nucleosynthesis was developed in the late 1940s by George Gamow, Hans Bethe (1906–2005), Robert Herman (1914–97) and Ralph Alpher (1921–2007). As a by-product of their work, they predicted the existence of the cosmic microwave background (CMB), but missed out on getting the credit when the CMB's discoverers, Robert Wilson (b.1936) and Arno Penzias (b.1933), won the Nobel Prize in 1978.

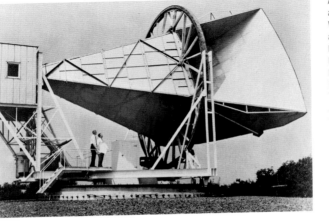

Arno Penzias and Robert Wilson standing at the Holmdel Horn Antenna at Bell Telephone Laboratories in Holmdel Township, New Jersey, USA, that brought their most notable discovery.

Forging the Elements

T =
20 min

KEY SCIENTISTS: GEORGE GAMOW • HANS BETHE • ROBERT HERMAN
RALPH ALPHER

Today, 92 naturally occurring elements make up the cosmos. However, in the first few moments of the Universe, there were just seven different atomic nuclei making isotopes of hydrogen, helium, lithium and beryllium – nothing else. An isotope is a variant of an element with the same number of protons as the basic form but a different number of neutrons in its nucleus.

These elements were forged during the first 20 minutes of the Universe. The young cosmos was dominated by nuclei of hydrogen, deuterium, helium-3, helium-4 and a small amount of lithium-7, along with traces of the unstable tritium (the isotope hydrogen-3) and beryllium-7. The dense, radiation-frazzled conditions were hot enough that nuclear fusion could take place to form these elements, but not so hot that any stray, whizzing photons could break the deuterium nuclei apart.

Big Bang nucleosynthesis wasn't capable of forging particularly heavy elements: soon, even the tritium and beryllium-7 decayed into helium-3 and lithium-7 nuclei, leaving just isotopes of hydrogen, helium and lithium in the young Universe.

KEY DEVELOPMENT
Astrophysicist Fred Hoyle developed a rival theory to the Big Bang, called the Steady State Theory, which tried to argue that the Universe was eternal without a beginning or an end. However, among the many reasons that Steady State was disproven was the fact that it could not replicate the relative abundances of the elements seen in the Universe. Big Bang Nucleosynthesis could, and was another feather in the cap of the Big Bang theory.

The nucleosynthesis periodic table. There are 92 naturally occurring elements, all of which have different origins. Only hydrogen, helium and a bit of lithium formed in the Big Bang.

STELLAR NUCLEOSYNTHESIS **p.208**

Beginnings of Structure

T =
<379,000
yrs

KEY SCIENTISTS: WERNER HEISENBERG

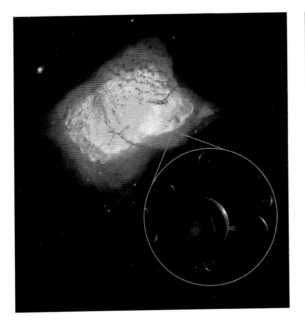

KEY DEVELOPMENT
In 1927 Werner Heisenberg
(1901–76) developed the
Uncertainty Principle. This
states that, at the quantum
level, matter and energy have
a wave-like quality, but even
if the value of one of these
properties is known, there is
no way of knowing the other.
If both are known to within a
degree of error, that error can
be no smaller than a quantity
known as Planck's constant
after German physicist Max
Planck (1858–1947).

Planetary nebula NGC 7027, in
which was detected helium hydride,
a combination of helium (red) and
hydrogen (blue): this was the first type of
molecule to form in the early Universe.

A key challenge of the Big Bang theory has been to explain how we got from a hot Big Bang of raw, seething energy to a Universe criss-crossed by a cosmic web of matter.

The early twentieth century saw physics and cosmology make huge advances in three directions that are relevant to the Universe: the development of Einstein's Theory of General Relativity, the observations of Edwin Hubble, and the quantum revolution that described the physics of the very small. This revolution was spearheaded by Einstein himself and others including Erwin Schrödinger (1887–1961), Niels Bohr (1885–1962), Werner Heisenberg and Paul Dirac (1902–84).

Since the Big Bang singularity was very small, quantum physics effects came into play. Quantum fluctuations – ripples of energy flashing in and out of existence – were expanded to huge scales by inflation and made permanent, and when the Universe's raw energy condensed into matter as the expanding cosmos cooled, these scaled-up quantum fluctuations ('baryonic acoustic oscillations') rippled through the primordial ocean of atomic nuclei (made from baryons) like waves. As the Universe cooled further, these waves were frozen into place and, as sites of greater density, became the scaffolding of the cosmic web of matter.

FILAMENTS AND VOIDS **p.62** CLUSTERS AND SUPERCLUSTERS **p.64**

Light and Matter Separate

T = 379,000 yrs

KEY SCIENTISTS: ROBERT WILSON • ARNO PENZIAS • JOHN MATHER
GEORGE SMOOT

The biggest evidence for the Big Bang comes in the form of Cosmic Microwave Background (CMB) radiation. Described as the leftover heat from the Big Bang fireball, it was predicted to exist by George Gamow, Ralph Alpher and Robert Herman, and discovered by Robert Wilson and Arno Penzias in 1964.

It represents what cosmologists term 'the moment of last scattering'. Before about 379,000 years after the Big Bang, the Universe was so hot that electrons could not bind to the atomic nuclei of hydrogen, helium and lithium that made up all the elements in the Universe at that time, to form complete atoms. Photons of light could not travel very far because they kept scattering off the free electrons.

Then at 379,000 years, the temperature of the Universe dropped below about 2,700 degrees Celsius – cool enough for electrons to bind to nuclei to form atoms. With the

electrons gone, the photons carrying the heat of the Big Bang were able to travel out into space freely, as light finally separated from matter. Those photons, having cooled to just 2.73 degrees above absolute zero, form the CMB.

KEY DEVELOPMENT

Arno Penzias and Robert Wilson, who were two astronomers working at Bell Labs in New Jersey, USA, discovered the CMB by accident. They were using a radio telescope called the Holmdel Horn Antenna when they detected an annoying background hiss at microwave frequencies that interfered with their observations. It was only when other astronomers, who were looking for the CMB, heard about it that the truth of the hiss was revealed.

The cosmic microwave background (CMB) radiation. The (false) colours indicate slightly warmer and colder regions, corresponding to areas of differing density in the early Universe.

The Cosmic Dark Age

T = <1 billion yrs

KEY SCIENTISTS: MARTIN REES • ABRAHAM LOEB • VOLKER BROMM

After the moment of last scattering, when the Universe had cooled enough for all the free electrons to be soaked up by atoms, light was free to travel through the cosmos unhindered. However, the dim glow of the CMB apart, there was nothing substantial to illuminate the inky blackness – no stars and no galaxies.

It was a gloomy time, which cosmologists call the Cosmic Dark Ages. The Universe was filled by neutral hydrogen and helium gas, and invisible dark matter. It took time for the Universe to be in position to cook up the very first stars and galaxies; before it could do so, the conditions had to be just right.

The moment of last scattering meant that there was no longer a cosmic ocean of plasma for the baryonic acoustic oscillations to ripple through, and they became frozen in place as regions of slightly higher density. These became the infrastructure of the cosmic web. Their gravity pulled in more dark matter and gas, the dark matter growing into haloes – roundish blobs with masses up to 100,000 times that of the Sun – containing enormous clouds of hydrogen atoms. It was these clouds that formed the first stars to begin the end of the Dark Ages.

KEY DEVELOPMENT

One of the big puzzles of cosmology is what played the biggest part in bringing the Dark Ages to an end – was it stars, or quasars, which are active black holes in the centres of galaxies that release huge amounts of light and radiation? The earliest quasar discovered so far existed 780 million years after the Big Bang.

Denser regions of the cosmic web drew in gas during the Cosmic Dark Ages. The massive gas clouds underwent gravitational collapse to form the first stars, ending the lightless period.

FILAMENTS AND VOIDS **p.62** DARK MATTER **p.214**

The First Stars

T = 180 million yrs

KEY SCIENTISTS: ABRAHAM LOEB • RICHARD LARSON • LEONARD SEARLE
EMMA CHAPMAN

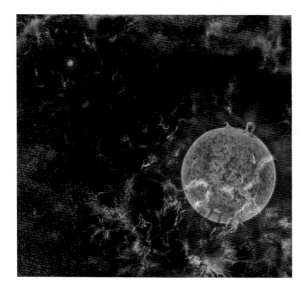

KEY DEVELOPMENT
The light from the very first stars is too faint to be seen directly by a telescope, but in 2018 astronomers detected a dip in the intensity of the CMB. This dip is caused by the ultraviolet light of the first stars exciting hydrogen gas so it can absorb some of the CMB photons. The dip is dated to 180 million years after the Big Bang.

An artist's impression of one of the first stars forming out of cosmic darkness.

By the time the Universe was 100 million–200 million years old, it was a place filled with mainly hydrogen and helium. Dense knots in what would gradually evolve into the cosmic web – the network of filaments that are the infrastructure of matter on the largest scales in the cosmos – allowed some of the very first structures to take hold – in particular, the first stars.

The first stars were composed of hydrogen and helium, unlike stars born today that are laced with other elements. But what they lacked in diverse composition, they made up for in size. These megasuns reached gigantic proportions, dwarfing the mass of our Sun by hundreds or even thousands of times, and were extremely hot, with surface temperatures of about 100,000 degrees Celsius. They were also exceedingly bright, lighting up the cosmic dark ages with the energy of a million Suns.

But much of that light wouldn't have been visible – given their temperature, it's likely that these ancient stars would have bathed their surroundings in ultraviolet light, lifting the neutral hydrogen fog that had descended upon the early Universe. This process is known as reionization, where the ultraviolet light gives the electrons in atoms enough energy to tear themselves from the atoms. Since electrons are negatively charged, losing them gave those atoms a positive charge.

STAR-FORMING NEBULAE **p.90**

The Death of the Megasuns

T = 200–300 million yrs

KEY SCIENTISTS: STAN WOOSLEY • VOLKER BROMM • MARTIN REES
HOWARD BOND

The first stars were probably even more massive than the most massive stars today, and would have died in titanic supernovae explosions.

With their gigantic size and brilliance, the first stars lived life with gusto, but that life didn't last long. They burned through their hydrogen at a rapid rate, lasting a million years or so (compared to the 10-billion-year lifetime of the Sun). When they reached the end, these stars went out in a blaze of glory.

Because they lacked any of the heavier elements, when they exploded they would have released energies 100 times more powerful than today's most energetic supernovae. These cataclysmic explosions would have resulted in two crucial consequences for the subsequent evolution of the Universe. One is that they produced a chemical footprint of newly created elements not formed in the Big Bang; these would be hoovered up by the next generation of stars. Second, they would have left behind monstrous, high-gravity black holes that could have been the seed of what would later be found at the centre of each large galaxy: a supermassive black hole weighing in at millions to billions of times the mass of our Sun.

KEY DEVELOPMENT
Nobody has ever seen one of the first stars or their supernovae explosions, but astronomers have detected the second generation of stars – some of which survive to this day, such as the star HE0107-5240, which is deficient in heavy elements. This indicates that it formed early, before the interstellar medium had been enriched by multiple generations of stars.

SUPERNOVAE **p.116** STELLAR BLACK HOLES **p.122**

Primeval Galaxies

KEY SCIENTISTS: GARTH ILLINGWORTH • RYCHARD BOUWENS
VOLKER BROMM • LEONARD SEARLE

As more and more stars were born, the first galaxies began to grow. However, there's a paradox at the heart of the primeval galaxies. At the centre of most large galaxies is a supermassive black hole, millions or billions of times the mass of the Sun. Astronomers have found that the mass of a supermassive black hole is proportional to the mass of the bulge of the galaxy that the black hole is found in. This tells us that the birth of the black hole and the galaxy are related, but which came first?

Astronomers are still trying to understand how supermassive black holes formed. One possibility is that they were created when lots of stellar mass black holes left by the supernovae of the first stars merged. However, supermassive black holes have been found in galaxies less than a billion years after the Big Bang – surely too soon for all those smaller black holes to merge. The alternative is that these supermassive black holes formed directly from the collapse of a giant cloud of gas at the centre of a burgeoning galaxy.

KEY DEVELOPMENT

In 2003–04, the Hubble Space Telescope spent a total of almost 12 days staring at a small patch of sky, in which it observed 10,000 distant galaxies. This 'Hubble Ultra Deep Field' was added to in 2009 and then again in 2012 (referred to as the eXtreme Deep Field). The oldest galaxies visible in the image date back to 600 million years after the Big Bang.

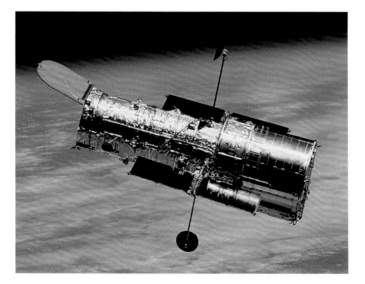

The Hubble Telescope in 2002, when spacecrews installed solar arrays, a camera, a power control unit, a reaction wheel assembly and an experimental cooling system. The telescope was first launched and deployed in 1990.

SPIRAL GALAXIES **p.66** GALAXY EVOLUTION **p.212**

Merging Galaxies

T = < 1 billion yrs

KEY SCIENTISTS: JENNIFER LOTZ • LEONARD SEARLE • WALLACE SARGENT

The first galaxies merged to form even bigger galaxies, eventually growing into giants far larger than our own Milky Way galaxy.

We see galaxies crashing together in the Universe today – titanic collisions that disturb the gas in those galaxies, prompting them to form new stars in a so-called starburst, or to feed the supermassive black hole at their heart, while gravitational tidal forces distort the shapes of those galaxies. Back when our cosmos was young, the Universe was smaller and the galaxies were much closer together, meaning that clashes between them were much more common.

The earliest known example of a galaxy merger has been seen by the Atacama Large Millimeter/submillimeter Array (ALMA) – a collection of 66 radio telescopes in the deserts of northern Chile. The particular merger is dubbed B14-65666 and appears 800 million years after the Big Bang.

Follow-up observations by the Hubble Space Telescope show that the galaxies are modest compared to our Milky Way, with a total mass of just 10 per cent that of our Sun. The two galaxies, which are merging to form a single galaxy, are bristling with young stars born in huge pockets of star formation.

KEY DEVELOPMENT

An early theory from Allan Sandage (1926–2010), Donald Lynden-Bell (1935–2018) and Olin Eggen (1919–98) in 1962 proposed that galaxies form from the collapse of a massive gas cloud, but this theory didn't explain all the features of galaxies. It was 16 years later that Leonard Searle (1930–2010) came up with the idea that galaxies may have formed from the bottom up, with smaller pieces merging to form a larger galaxy.

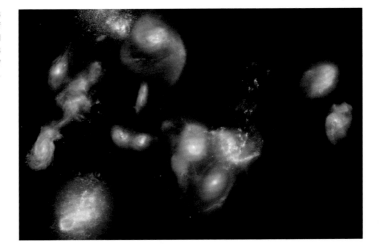

An artist's impression of colliding and merging galaxies in the early Universe.

SPIRAL GALAXIES **p.66** ELLIPTICAL GALAXIES **p.68** DWARF GALAXIES **p.72**

Birth of the Milky Way

KEY SCIENTISTS: DIEDERIK KRUIJSSEN • CRISTINA CHIAPPINI
GERRY GILMORE • ROSEMARY WISE

In this artistic impression, we see our Milky Way being born. There are no spiral arms, just an agglomeration of star clusters surrounding an active supermassive black hole.

The mergers of smaller galaxies to form larger galaxies is called hierarchical merging. Our Milky Way formed in just this way. The Milky Way is a spiral galaxy – it has a large disc where the stars and gas are arranged in arms that appear to be spiralling in. At the centre of the disc is the bulge, which houses a supermassive black hole. Surrounding the disc and centred on the bulge is the halo – a roughly spherical cloud of stars.

The bulge and the halo are the oldest parts of the galaxy, and would have formed first, 13 billion years ago, from the mergers of small dwarf galaxies, compact and dense like oversized versions of the globular star clusters found in the halo today. The spiral disc is younger, less than 10 billion years old, and also would have grown by capturing smaller galaxies. Two of the more notable collisions were with a huge galaxy known as Kraken and a sausage-shaped structure dubbed Gaia-Enceladus.

KEY DEVELOPMENT

In 2020, astronomers were able to disentangle the Milky Way's merger history, showing that the Milky Way has experienced about 20 mergers during its history, with the most massive taking place between 6 billion and 10 billion years ago – the largest being with the Kraken Galaxy, which dramatically increased the size of the Milky Way.

THE MILKY WAY **p.84** THE HALO AND GLOBULAR STAR CLUSTERS **p.86** SPIRAL ARMS **p.88**

Cosmic Recycling

Ongoing

KEY SCIENTISTS: FRED HOYLE • MARGARET BURBIDGE
GEOFFREY BURBIDGE • WILLIAM FOWLER

There's an adage that stars like to pollute while the galaxies clean up their mess. The stars explode in their demise, or expand into red giants and planetary nebulae, either way spewing out a range of elements that these hot furnaces had manufactured in their core.

The first stars were made entirely of hydrogen and helium. When they died, they released heavier elements that were then recycled into the next generation of stars. When this second generation died, they added new elements, and on and on, each generation further enriching the inter-stellar medium with the different elements of the Periodic Table. In essence, galaxies are set up like recycling centres, ensuring that the next generation of stars is always being born from the material their parent stars left behind.

When the first stars formed, there weren't the heavy elements present to build rocky planets and life. Only the elements formed inside stars and released in death could breathe life into these worlds. As Carl Sagan (1934–96) once famously said, we are made of 'starstuff'.

KEY DEVELOPMENT
One of the oldest planets known in the Universe is TOI 561, discovered by NASA's Transiting Exoplanet Survey Satellite (TESS). Its age is 10 billion years, and it is one and a half times as big as Earth, thought little else is known about it. Its existence proves that there were at least enough heavy elements to be building rocky planets less than 4 billion years after the Big Bang.

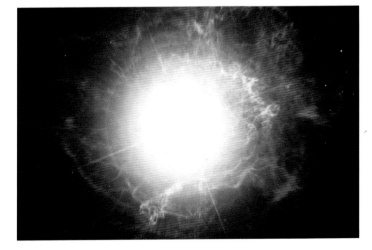

When massive stars explode as supernovae, such as that depicted in this artwork, they spread the heavy elements formed within the stars into space.

STELLAR NUCLEOSYNTHESIS **p.208**

Origins of the Solar System

T = 9.2 billion yrs

KEY SCIENTISTS: PIERRE-SIMON LAPLACE • ALAN BOSS

Over four-and-a-half billion years ago, our Solar System began to swirl into existence out of a gigantic cloud of molecular gas and dust that was undergoing gravitational collapse. This cloud was 98 per cent hydrogen and helium, and 2 per cent heavy elements produced by generations of previous stars, and now being recycled. At the centre of this inflating cloud the Sun started to form, while around it the gas cloud flattened into a spinning disc inside which the planets formed.

Originally, the cloud that gave birth to the Sun may have measured 65 light years across. The cloud broke up into numerous fragments, each fragment the beginnings of a new star and planets. Our fragment – the 'pre-solar nebula' – contained a total mass of a little over the mass of our Sun.

As it collapsed, the nebula spun faster and faster, and its core grew denser and denser as more gas fell onto it from the solar nebula. When 100,000 years had passed, a concoction of gravity, magnetic fields and the pressures of gas and rotation had flattened it into a disc measuring roughly 200 astronomical units – 29.9 billion kilometres (18.6 billion miles) – across.

KEY DEVELOPMENT

In 2020, scientists analysed pre-solar grains of silicon carbide found in a meteorite and discovered that some of them were up to 7 billion years old, formed by stars that lived long before our Solar System. The discovery is a great example of cosmic recycling, and how elements produced in one generation of stars wind up in the next.

The Solar System formed from a spinning disc of gas and dust around the young Sun, as seen in this artist's impression.

METEORITES **p.142** COLLISIONAL ACCRETION (SOLAR SYSTEM FORMATION) **p.199**

The Sun Ignites

T = 9.2 billion yrs

KEY SCIENTISTS: GEORGE HERBIG • GUILLERMO HARO • ALFRED JOY

KEY DEVELOPMENT

The molecular cloud that fragmented and gave birth to the Sun should also have formed many other stars, and astronomers are eager to find these solar siblings. It is thought that two have been identified so far: HD 162826, which is 110 light years away and slightly hotter than the Sun, and HD 186302, which is now 184 light years away and is almost exactly like the Sun.

We've got gravity to thank for igniting our Sun. It was this force that pulled material to the centre of the rotating protoplanetary disc, making it not just super-hot, but also a highly pressurized environment.

When you crush hydrogen atoms under immense temperature and pressure, they combine with each other – fuse – to form helium. In the process, energy is released, and with this first burst of energy our Sun shone into existence.

It took a while for the young Sun to reach this stage. Before its core grew hot and dense enough for nuclear fusion reactions, it shone thanks to energy released as a result of its gravitational contraction. At this stage, still less than 10 million years old, our young Sun was what we call a T Tauri star – named after a young star that resides in the Taurus Molecular Cloud and which is characterized by bipolar jets, produced when inflating matter is beamed away by magnetic fields.

Just before it reached its 50 millionth birthday, our Sun's core temperature reached 15 million degrees Celsius, and it began its life as a fully-fledged star. It slowly began leaving behind its stellar siblings, which were born from the same cosmic cradle of gas and dust.

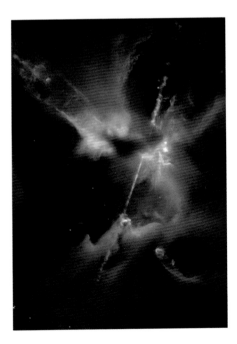

Jets are ejected from the poles of a young star forming inside a nebula of gas in this Hubble Space Telescope picture.

THE SUN **p.126**

Creation of the Planets

KEY SCIENTISTS: ALAN BOSS • JACK LISSAUER • HAL LEVISON

The planets were formed through a process of collision and accretion, building up from small pebbles into asteroid-sized bodies and then into protoplanets.

Our Sun didn't gobble up all of the matter when it formed, it left some in the guise of a protoplanetary disc that built the planets. As the disc cooled, gas condensed out into dust grains. Conditions close to the Sun were still warm, allowing only for metals and silicates to condense out. The rocky planets – Mercury, Venus, Earth and Mars – were assembled from this material, as were numerous other protoplanets, which have since been lost to time, ejected from the Solar System or destroyed.

Farther from the Sun, where it was colder, there was more gas. Very quickly, within 10 million years or less, the gas giants Jupiter and Saturn formed, the first planets in the Solar System to do so, as their huge mass was able to sweep up the gas quickly. Their gravitational influence would also have had an impact on the formation of the smaller inner planets. Meanwhile, beyond them, Uranus and Neptune formed last, in a region where volatiles such as water, nitrogen and carbon dioxide froze out into ices.

KEY DEVELOPMENT
A lot of what we know about how the Solar System formed comes from studies of exoplanets forming around other stars. The Atacama Large Millimeter/submillimeter Array (ALMA) in Chile is able to detect radio waves emitted by dust in protoplanetary discs, revealing the complexities of how planets are born.

THE SOLAR SYSTEM **p.124**

Birth of the Moon

T = 9.3 billion yrs

KEY SCIENTISTS: ROBIN CANUP • BILL HARTMANN • DONALD DAVIS
ALASTAIR CAMERON • WILLIAM WARD

The Moon formed from the debris of the collision between a Mars-sized protoplanet and the young Earth.

The making of our Moon was violent. An object named Theia, roughly the size of Mars, smashed into a young Earth some 100 million years after the Solar System began to form. The collision is thought to have had an energy around 100 million times larger than the impact that wiped out the dinosaurs. Computer simulations suggest that Theia crashed into Earth at a 45-degree angle at a speed of 4 kilometres (2.49 miles) per second.

In doing so, Theia tossed some of its crust and that of our planet's into outer space, where it formed a ring around Earth, while Theia's core and mantle combined with that of Earth. Over time, the molten material in that ring came together to form the Moon.

Although there have been several theories as to how our Moon was made, the giant-impact hypothesis is the most widely accepted, especially given supporting evidence, which includes how our Moon seems to be made mostly from material similar to that of Earth's crust.

KEY DEVELOPMENT
The giant impact theory behind the origin of the Moon was first developed in two academic papers of the 1970s, one by Bill Hartmann (b.1939), Donald Davis (b.1952) and Alfred Cameron (1925–2005), the other by William Ward (1944–2018). Then, at a conference of lunar scientists 10 years later, the theory emerged as the most popular, and has been advanced upon ever since, particularly thanks to the efforts of Robin Canup (b.1968) of the Southwest Research Institute in Boulder, Colorado.

THE MOON **p.138**

Order of the Planets

KEY SCIENTISTS: HAL LEVISON • ALESSANDRO MORBIDELLI
KEVIN WALSH

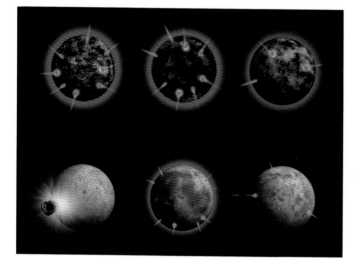

This artwork illustrates the bombardment history of the Moon, starting top left and ending bottom right. Note how periods of intense heat are interspersed with cooler interludes. Similar impacts have occurred throughout history on every planet.

Today our Solar System has an ordered set-up, with the planets remaining firmly in their orbits, but it wasn't always this way. Some shuffling – planetary migration – took place, changing their orbits.

The first theory of this turbulent time was the Nice model, named after the location of the Côte d'Azur Observatory in France where the idea was developed. It states that the giant planets – Jupiter, Saturn, Uranus and Neptune – formed twice as close to the Sun, compared to where they orbit now, and slowly migrated outwards, scattering in their wake asteroids and comets, many of which impacted the inner planets in an event known as the Late Heavy Bombardment (LHB).

Because evidence for the LHB doesn't stack up, astronomers have begun to consider another theory, the Grand Tack model, according to which Jupiter, just after its formation, began migrating inwards quickly, getting as close to the Sun as Mars is today, until Saturn caught up with it and its gravity helped to reverse Jupiter's course. Jupiter's migration would have had important consequences for how the inner planets, especially Mars, formed.

KEY DEVELOPMENT

In 1995, Michel Mayor (b.1942) and Didier Queloz (b.1966) discovered the first exoplanet around a Sun-like star. This planet, 51 Pegasi b, was a gas giant very close to its star. However, theories of how planets form state that gas giants cannot form so close to their star. Therefore, 51 Pegasi b and other worlds like it are proof that they must have migrated inwards from where they formed.

EXOPLANETS **p.94** JUPITER **p.154** PLANETARY MIGRATION **p.200**

Soaking up the Debris

KEY SCIENTISTS: GENE SHOEMAKER • EDWARD CHAO
LUIS AND WALTER ALVAREZ

The growing planets accreted more material by sweeping up the debris in the Solar System, such as comets and asteroids, which impact on their surface.

Any material that didn't coalesce to form planets in the dusty construction yard that was the young Solar System found a home elsewhere: in the Asteroid Belt between Mars and Jupiter, in the Kuiper Belt beyond the orbit of Neptune, or in the space between the planets.

Given the high melting points in the early inner Solar System, all of the millions of lumps of space rock that occupy the Asteroid Belt are rocky and infused with carbon, silicon, oxygen and metals. The Kuiper Belt is freezing, and the planetary detritus out there is made from ice and frozen compounds such as methane and ammonia.

The planets, or the cores of the gas giants, formed by colliding with these pieces of rubble, large and small. While that process may have slowed, it has not stopped. Craters on the Moon were formed by impacts. We occasionally spot a comet or small asteroid colliding with Jupiter, while Earth may have received its water from either comet or asteroid impacts, and an impact wiped out the dinosaurs.

KEY DEVELOPMENT
Earth's water has a very specific ratio of hydrogen to deuterium (an isotope of hydrogen), so the source of Earth's water must share that same ratio. The obvious source of that water was comets, but studies of multiple comets have shown that they have a different ratio. Water-rich asteroids are now the key suspects, with carbonaceous chondrite asteroids having almost exactly the same ratio as Earth's oceans.

THE ASTEROID BELT **p.148** THE KUIPER BELT AND ITS MEMBERS **p.184** COMETS **p.188**

The Development of the Sun

T = ~20 billion yrs

KEY SCIENTISTS: ARTHUR EDDINGTON • FRED HOYLE • JULIANA SACKMANN
KATHLEEN KRAEMER

At an age of 4.6 billion years, our Sun is halfway through its lifespan, and will spend a further 7 billion years burning hydrogen in its core. In that time, it's going to get hotter and brighter, its luminosity increasing by 6 per cent every billion years. That doesn't sound like a great deal, but it means that in about a billion years' time, the Sun's growing luminosity will render Earth too warm for life to survive. The oceans will evaporate, and our planet will become a scorched dust bowl.

Eventually, near the end of the Sun's life, that hydrogen is going to run out and from this point onwards, our star will begin to evolve into something new. In its core, the hydrogen will have all been fused into helium, so as nuclear reactions in the core stutter, the Sun's interior will contract under gravity, causing the temperature to rise so that hydrogen in a shell around the core will begin fusion reactions. This so-called shell burning will cause the Sun to swell to a radius a thousand times larger than its current dimensions, and become a red giant.

KEY DEVELOPMENT

In 1910, Ejnar Hertzsprung (1873–1967) and Henry Norris Russell (1877–1957) independently developed what today we call the Hertzsprung–Russell (HR) diagram, which plots stellar luminosity against colour (temperature). Different types of stars fall into different areas of the diagram, but it wasn't until after the discovery of hydrogen fusion in the 1940s that there was an understanding of how the HR diagram shows the evolution of stars, from hydrogen burning to red giants.

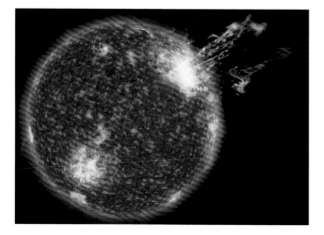

As our Sun ages, it will grow hotter and more luminous, making life on Earth impossible in about a billion years' time.

RED GIANTS **p.110** STELLAR EVOLUTION **p.207**

Future of the Solar System

T =
~20 billion
yrs

KEY SCIENTISTS: JULIANA SACKMANN • KATHLEEN KRAEMER
ARNOLD BOOTHROYD

As our Sun swells into a red giant, the planets in close orbit around it will be unable to get out of the way. Mercury and Venus will be engulfed in its swollen layers and destroyed. Eventually it will reach Earth, and quite possibly our planet will be swallowed up too.

Whether our Sun will gobble up Earth or not will depend on how big it does get, but either way, it's not good news for our planet. As the limb of the red giant inches closer, intense radiation and heat will turn Earth's surface molten before vaporizing it, while steadily stripping away the atmosphere. By the end, Earth would be a husk, nothing more than a dense, iron core. Mars may escape the Sun's reach, and for a time grow warm enough to support liquid water, but this too would eventually evaporate.

The icy moons of the outer planets may fare better. As the Sun brightens and grows, the habitable zone – the distance from the Sun where temperatures are just right for liquid water – will move outward, perhaps bringing life to ocean moons such as Titan, Europa and Enceladus.

KEY DEVELOPMENT
In 1997 scientists Ralph Lorenz (b.1969), Jonathan Lunine (b.1959) and Chris McKay (b.1954) showed that when the Sun turns into a red giant, Saturn's moon Titan – which today is a frigid body rich in water-ice and hydrocarbons that could form the building blocks of organic molecules for biology – could develop oceans and a wealth of organic compounds. However, this habitable period would only last a few hundred million years.

As the Sun expands into a red giant, it will swell in Earth's sky, swallowing Mercury and Venus and eventually our planet as well.

RED GIANTS **p.110** EUROPA **p.158**

Death of the Sun

T = ~20 billion yrs

KEY SCIENTISTS: WILLIAM HERSCHEL • ARTHUR EDDINGTON
HENRY NORRIS RUSSELL • EDWARD PICKERING • WILLIAMINA FLEMING

As the Sun evolves into a red giant, its core will continue to contract under gravity and its core temperature will rise to 100 million degrees Celsius (it is currently around 15 million degrees Celsius). At this temperature, fusion reactions of helium start to create carbon, and thus the Sun will get a second lease of life.

As it chomps through the helium in its core, the Sun's temperature continues to rise until it reaches 300 million degrees Celsius. This is hot enough to ignite all the core helium at once, causing a blast known as the helium flash. Subsequently, the helium burns in a shell around the core. These are the Sun's final days. The shell burning makes the Sun become unstable and start pulsing, and it takes a lot out of the Sun, particularly in the way of mass, some of which it loses with each pulse. The Sun's outer layers are puffed off to form a beautiful planetary nebula, while exposing its core, a remnant that takes the form of a white dwarf.

KEY DEVELOPMENT

Many planetary nebulae sport beautifully symmetrical shapes that resemble cosmic butterflies or jellyfish. Recent observations and modelling of the complex shapes of planetary nebulae suggest that they may be produced by binary star systems, as opposed to single stars like the Sun. This might mean that when the Sun dies, its planetary nebula will look rather boring, compared to the wonderfully twisting and corkscrewing shapes of other planetary nebulae.

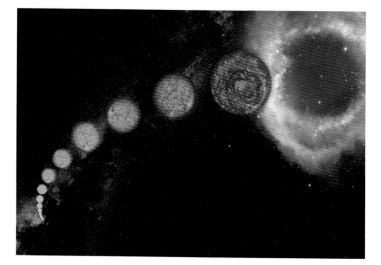

An artist's impression of the final phases of the Sun, which will grow in size as it loses power.

PLANETARY NEBULAE **p.112** WHITE DWARFS **p.114** STELLAR EVOLUTION **p.207**

The Far Future

T =
1
quadrillion
yrs

KEY SCIENTISTS: FRED ADAMS • GREGORY LAUGHLIN • ABRAHAM LOEB

An artist's impression of the Andromeda Galaxy nearing the Milky Way.

The deep future will see everything change in the heavens. The stars may seem fairly constant to us, but just like our Sun they will all eventually die, either by becoming red giants or exploding as supernovae. However, for all the stars that are lost, there will be many more new ones.

Some 4 billion years from now, the Andromeda Galaxy and its trillion stars will crash into our Milky Way and merge to form a new elliptical galaxy, which astronomers have named 'Milkomeda'. No stars will actually collide during the merger, but huge gas clouds will, instigating a new, energetic round of star formation.

Wait long enough and every star will die, and all the gas in the Universe will be used up so no new stars can form. All that will be left will be white dwarfs, black holes and neutron stars. The white dwarfs will slowly cool to become black dwarfs – frozen remnants that won't emit any light or heat. It will take our Sun at least a quadrillion years to evolve to this stage.

KEY DEVELOPMENT
The merger of the Milky Way and the Andromeda Galaxy will not be the end of our galaxy's evolution. The other galaxies in our cluster, the Local Group, will also eventually collide and merge with Milkomeda. Meanwhile, in about 100 billion–150 billion years' time, cosmic expansion will have carried all the other galaxies so far away from Milkomeda that the future inhabitants of that galaxy will no longer be able to see any other galaxies in the night sky.

GALAXY EVOLUTION **p.212**

The Fate of the Universe

T = >10^{100} yrs

KEY SCIENTISTS: JOHN BARROW • FRANK TIPLER • MARTIN REES STEPHEN HAWKING

The deciding factor in how the Universe will end depends on the behaviour of dark energy, which is accelerating the expansion of the cosmos.

Much depends on whether this energy weakens or gets stronger over time. If it weakens, it may cause gravity to lead the Universe slowly to contract back on itself in a Big Crunch.

If the dark energy gets stronger, or at least remains the same, then there are two possible scenarios, both of which involve the Universe expanding forever, until galaxies are carried so far from one another that they disappear over the cosmic horizon. One possibility is that the expansion will slow but never completely stop, leaving just a big, dead void that undergoes 'heat death' – the entropic decay of all matter, including eventually subatomic particles such as protons, into radiation. Alternatively, if dark energy grows much stronger, it could tear the fabric of spacetime apart in a Big Rip.

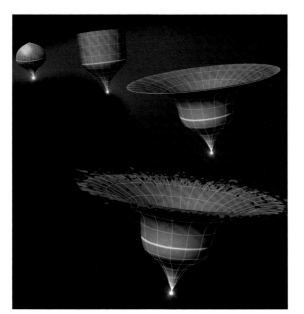

KEY DEVELOPMENT

The fate of the Universe rests on dark energy, but we don't yet know what dark energy is. If it does resemble Einstein's cosmological constant, then it will only grow stronger in time as new space is created in the expansion. However, if it is a quantum 'scalar' field known as quintessence, then its strength could vary with space and time, leaving the fate of the Universe uncertain.

Possible fates for the Universe. Top left is a Big Crunch collapse; to its right are the two scenarios in which it expands forever, one diverging more slowly than the other; the bottom illustration is of a Universe that accelerates as it expands until it tears itself apart in a Big Rip.

MULTIVERSE **p.198** DARK ENERGY **p.215**

Components

Filaments and Voids

PATTERNS IN THE DISTRIBUTION OF GALAXIES THAT FORM THE LARGEST COSMIC STRUCTURES

The very largest structures in the Universe are extended strands and thin sheets within which matter is concentrated. Known as filaments, they extend for hundreds of millions of light years, merging and connecting with each other at the edges to surround vast, apparently empty spaces known as voids. The filaments are rendered visible by the light and other radiations from countless galaxies held together by gravity in clusters and superclusters.

Considering that the visible matter in galaxies accounts for only one sixth of all the mass in the Universe, the vast and apparently empty voids might seem like ideal places for the unseen dark matter to lurk – but measurements of large-scale motion among the clusters and superclusters show no great signs that they are being affected by the gravitational pull of the voids (as could be expected if they contained dark matter).

Indeed, these structures are so vast that they cannot be explained as concentrations of matter pulling together under the influence of gravity – there simply hasn't been enough time in the 13.8 billion years since the Big Bang for such large structures to form. Instead, it seems that the web-like structure of the Universe on the largest scales is an echo of creation itself – the result of variations in the distribution of matter very shortly after the Big Bang.

MARGARET GELLER
Margaret Geller (b.1947) led some of the first attempts to map large-scale cosmic structure. The galaxy redshift survey carried out at Harvard-Smithsonian Center for Astrophysics in the 1980s measured galaxy red shifts (a proxy for distance) across a broad slice of the sky, and led to the discovery of the prominent nearby filament known as the Great Wall.

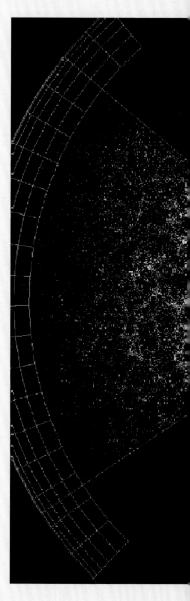

A redshift survey, carried out by the 3.9m Anglo-Australian Telescope, which reveals the Universe's large-scale structure to a depth of 2.5 billion light years.

DISTRIBUTION OF MATTER **p.16** BOUND SYSTEMS **p.18** BEGINNINGS OF STRUCTURE **p.40**
THE COSMIC DARK AGE **p.42**

Notable filaments
CfA2 Great Wall (Coma Wall)
Sloan Great Wall (Corvus-Hydra-Centaurus Wall)
Sculptor Wall
Hercules-Corona Borealis Great Wall

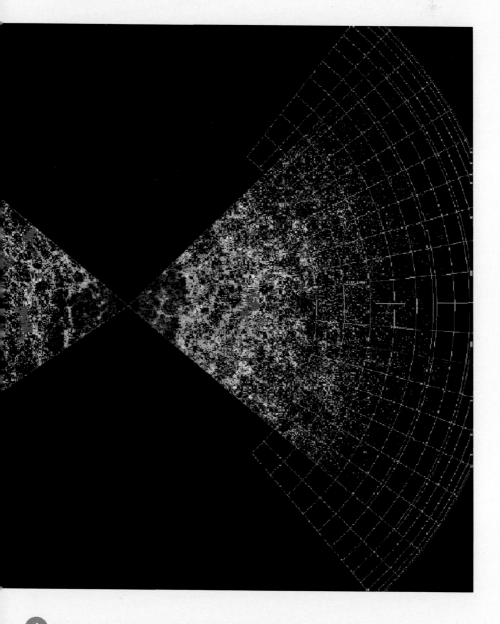

UNIVERSAL GRAVITATION **p.201** GALAXY EVOLUTION **p.212** DARK MATTER **p.214** DARK ENERGY **p.215**

Clusters and Superclusters

GROUPS OF GALAXIES BOUND TOGETHER BY THEIR COLLECTIVE GRAVITATIONAL PULL

Galaxy superclusters
Laniakea supercluster, Virgo
Southern supercluster, Fornax
Saraswati supercluster, Pisces
Coma supercluster, Coma

While stars and planets are typically separated by gulfs of empty space that are vast compared to their own size, galaxies are – relatively speaking – far more crowded together. At least in the local Universe, about 80 per cent of known galaxies are gravitationally bound in dense clusters or relatively sparse 'groups' (such as the Milky Way's own 'Local Group'). The number and mass of galaxies within these structures can vary from a few dozen to many hundreds, but they typically occupy a similar volume of space – perhaps 10 million light years across. The motion of galaxies within clusters suggests they are being affected not only by the gravitational pull of their visible companion galaxies, but also by the influence of vast concentrations of unseen dark matter accounting for the majority of a cluster's mass.

Clusters typically blend into one another around their fringes, giving rise to vast superclusters perhaps a few tens of millions of light years across. All the galaxies within a supercluster show signs of moving under the gravitational influence of their common centre of mass, making superclusters the largest 'gravitationally bound' structures in the cosmos and suggesting that while their locations are governed by variations in matter density after the Big Bang, their actual structure is mostly the result of gravity acting over billions of years.

GEORGE ABELL
George Ogden Abell (1927–83) played a leading role in the collection and analysis of data for the Palomar Observatory Sky Survey. This long-exposure photographic survey captured millions of faint galaxies for the first time and showed conclusively that many were concentrated in dense clusters, of which Abell catalogued more than 2,700.

DISTRIBUTION OF MATTER **p.16** BOUND SYSTEMS **p.18** BEGINNINGS OF STRUCTURE **p.40**
THE COSMIC DARK AGE **p.42**

Abell 2744 YI, a galaxy in the cluster Abell 2744, which is smaller than the Milky Way but produces more stars.

UNIVERSAL GRAVITATION **p.201**   GALAXY EVOLUTION **p.212**   DARK MATTER **p.214**

Spiral Galaxies

DISC-SHAPED STAR SYSTEMS WITH ABUNDANT STAR FORMATIONS IN SPIRAL ARMS

Spirals are the most spectacular galaxies – vast cosmic pinwheels packed with stars, dust and gas. They vary in size between a few tens of thousands and perhaps 200,000 light years in diameter, and show considerable differences in structure. Classic spirals have a densely packed central hub of older red and yellow stars, shaped like a slightly squashed bun and surrounded by a thinner disc of orbiting stars. Two or more arms dominated by bright young blue stars and white stars emerge from the hub and trace a spiral path across a flattened disc of varied, middle-aged stars – their shape and definition help to define different categories of spiral. Barred spirals (which include our own Milky Way galaxy) have a central bar of older stars that crosses the hub, with the spiral arms rooted to either end.

Despite first appearances, galaxy spiral arms are not physical chains of bright stars (which would wind up and be pulled into the hub over a few galactic rotations) but are instead regions of greater density and increased star-formation. Almost all stars are born in spiral arms – the heaviest and brightest live and die before their orbits carry them elsewhere, while longer-lived, smaller and fainter stars survive to move into the general stellar disc.

WILLIAM PARSONS
Anglo-Irish nobleman William Parsons, 3rd Earl of Rosse (1800–67) built the world's largest telescope (with a mirror diameter of 1.83 metres/72 inches) at Birr Castle in Ireland. While studying the mysterious fuzzy objects then known by the catch-all of 'nebulae', he was the first person to note that some of them showed a spiral structure.

Notable spiral galaxies
Andromeda Galaxy (Messier 31), Andromeda
Triangulum Galaxy (M33), Triangulum
Bode's Galaxy (M81), Ursa Major
Southern Pinwheel Galaxy (M83), Hydra
Whirlpool Galaxy (M51), Canes Venatici

BOUND SYSTEMS **p.18** MERGING GALAXIES **p.46** BIRTH OF THE MILKY WAY **p.47**

The spiral galaxy NGC 3147 is around 140,000 light years
across and around 130,000 light years from Earth.

 UNIVERSAL GRAVITATION **p.201** STAR FORMATION **p.209** DENSITY WAVES AND GALAXY STRUCTURE **p.211**
GALAXY EVOLUTION **p.212**

Elliptical Galaxies

SPHERICAL OR ELONGATED BALLS OF ANCIENT RED AND YELLOW STARS

CHARLES MESSIER
Many of the sky's brightest star clusters, galaxies and nebulae were first systematically catalogued by French astronomer Charles Messier (1730–1817). Messier's sky survey ultimately listed 102 objects, including the giant ellipticals M49 and M87, and the smaller M32, a satellite of the Andromeda spiral galaxy. Later astronomers expanded the list to 110.

Ball-shaped or elliptical galaxies account for about 20 per cent of all known galaxies. They encompass both the smallest and largest galaxies, and vary in shape from densely packed spheres and elongated cigars to loose, sparsely populated balls. Mostly lacking in the raw materials of star formation, ellipticals are composed almost entirely of older red and yellow stars. These follow elliptical orbits around a shared centre of gravity in the core of the galaxy, but their orbits are also tilted at a variety of angles, giving rise to a ball-shaped appearance.

Ellipticals are thought to form when other galaxies collide and merge. During these cataclysmic events, collisions between individual stars are rare, but star-forming gas clouds slam into each other directly, heating up to a point where they can escape the galaxy's gravity and 'boil away' into surrounding space. So although the process initially triggers waves of star formation, once this comes to an end and the brightest stars age and burn out, the galaxy is left with only its longer-lived, fainter and cooler stars. Lenticular galaxies, which have an elliptical hub surrounded by a disc of gas and stars but no ongoing star formation, may represent a stage in the aftermath of such collisions.

Notable elliptical galaxies
Messier 32 (dwarf), Andromeda
Maffei 1, Cassiopeia
NGC 5128 (lenticular), Centaurus
Messier 49 (giant), Virgo
Messier 87 (giant), Virgo

BOUND SYSTEMS **p.18** MERGING GALAXIES **p.46**

ESO 325-G004 is an elliptical galaxy 416 million light years
from Earth in the Centaurus constellation.

UNIVERSAL GRAVITATION **p.201** GALAXY EVOLUTION **p.212** ACTIVE GALACTIC NUCLEI **p.213**

Irregular Galaxies

CLOUDS OF GAS, DUST AND YOUNG, BRIGHT STARS

Notable irregular galaxies
Large Magellanic Cloud, Dorado/Mensa
Small Magellanic Cloud, Tucana
NGC 2337, Lynx
NGC 1427A, Fornax

BOUND SYSTEMS **p.18**   PRIMEVAL GALAXIES **p.45**   MERGING GALAXIES **p.46**

HENRIETTA SWAN LEAVITT

Henrietta Swan Leavitt (1868–1921) is famous for her study of the Small Magellanic Cloud. Reasoning that stars in this isolated clump are effectively at the same vast distance from Earth, she treated their apparent brightness as a proxy for their true luminosity. This revealed a link between the luminosity and pulsation period in variable stars named Cepheids – a relationship that proved vital to measuring cosmic distances.

So-called irregular galaxies are somewhat shapeless clouds of stars, gas and dust that often show plentiful signs of ongoing star formation. Although smaller than spirals, they can be anything from a few thousand to a few tens of thousands of light years across, and account for about a quarter of all galaxies in the modern Universe. Observations of distant space suggest they were much more abundant in the Universe's first few billion years.

Irregulars are broadly divided into two simple categories. 'Irr-I' galaxies show faint but detectable signs of structure such as central bars and loose, faint spiral arms, while 'Irr-II' galaxies are entirely shapeless. The most famous examples are the Large and Small Magellanic Clouds – substantial satellites of our own galaxy that appear as detached portions of the Milky Way in far southern skies, both of which fall into the Irr-I category.

In general, it seems that spiral structure begins to emerge inevitably as gas, dust and stars interact with each other and transform larger irregular galaxies into flattened rotating discs. The higher frequency of small irregulars in the distant past is an indication that they are galaxy building blocks whose frequent mergers slowly built the large spiral galaxies we see in today's Universe.

Henrietta Swan Leavitt working at her desk at the Harvard College Observatory in around 1910.

UNIVERSAL GRAVITATION **p.201** STAR FORMATION **p.209** DENSITY WAVES AND GALAXY STRUCTURE **p.211**
GALAXY EVOLUTION **p.212**

Dwarf Galaxies

THE SMALLEST AND FAINTEST CLOUDS OF STARS

In between the bright, large galaxies of the Universe lie smaller systems that are normally dim and diffuse. The faintness of these galaxies means they are hard to detect beyond our nearby region of space, but the Milky Way and Andromeda galaxies each have retinues of a couple of dozen such galaxies, and dwarfs are increasingly being discovered in more distant clusters. Dwarf galaxies display most of the familiar galaxy shapes – including elliptical, irregular and even spiral structures – plus some unusual forms that are never found in larger systems. These include the dwarf spheroidals (loose spherical clouds, populated with small numbers of older stars and very little dust) and ultra-compact dwarfs – small but incredibly crowded star clouds that may be the exposed cores of elliptical galaxies that have had their outer layers stripped away.

Despite their relatively small numbers of stars, dwarf galaxies seem able to resist being torn apart by the stronger gravity of other galaxies during close encounters. This suggests that they have a large gravitational anchor of their own, which holds them together – most likely a central black hole that may account for more than 10 per cent of a dwarf galaxy's entire mass.

HARLOW SHAPLEY
The first dwarf spheroidal galaxy was discovered in the constellation of Sculptor by Harlow Shapley (1885–1972). Shapley initially mistook it for a distant galaxy cluster before determining that the individual scattered members were faint stars, and concluding that this was a new star system orbiting the Milky Way.

Notable dwarf galaxies
SagDEG (dwarf spheroidal), Sagittarius
Canis Major Overdensity, Canis Major
NGC 147 (dwarf spheroidal), Cassiopeia
NGC 185 (dwarf spheroidal), Cassiopeia
SagDIG (dwarf irregular), Sagittarius

BOUND SYSTEMS **p.18** PRIMEVAL GALAXIES **p.45** MERGING GALAXIES **p.46**

Harlow Shapley's early ambition was to be a journalist,
but when the course he wanted was postponed he chose
instead to study astronomy.

UNIVERSAL GRAVITATION **p.201** GALAXY EVOLUTION **p.212**

Interacting Galaxies

CLOSE ENCOUNTERS AND COLLISIONS THAT DRIVE GALAXY EVOLUTION

Alongside the major galaxy types is a wide range of systems that have generally been classified as 'peculiar'. They typically show some but not all features of at least one of the major galaxy types – for instance, displaying more than one hub-like core of red and yellow stars, elongated 'streamers' instead of spiral arms, or even wheel-like disc or spoke-like structures.

Most of these curious galaxies can be explained by collisions and interactions between individual galaxies. Such events are widespread in the Universe – galaxies are typically separated by gaps on a similar scale to the galaxies themselves, and their enormous mass allows them to extend a powerful gravitational attraction over their surroundings. Even if galaxies do not collide directly, the tidal forces they raise – a result of the difference in one galaxy's gravitational pull on the near and far sides of the other – can disrupt the normal orderly orbits of stars and other material, causing spiral arms to unwind and compressing gas and dust to produce huge regions of violent star formation. On a smaller scale, close encounters in which a smaller dwarf galaxy passes close to a larger one seem to temporarily intensify the rate of star formation in spiral arms – though ultimately the dwarf galaxy may be torn apart, with its stars and other materials absorbed into the larger system.

HALTON ARP
During the 1960s, Halton Arp (1927–2013) compiled the first Atlas of Peculiar Galaxies, believing that these strange galactic misfits could hold the key to explaining how galaxies evolved. While the work itself proved valuable, Arp's own belief – that these galaxies are caught in the process of expelling, rather than capturing, material – was eventually disproved.

BOUND SYSTEMS **p.18** PRIMEVAL GALAXIES **p.45** MERGING GALAXIES **p.46**

Notable interacting galaxies
Messier 51/NGC 5195, Canes Venatici
'The Mice', NGC 4676, Coma Berenices
Arp 87, NGC 3808/3808A, Leo
Stephan's Quintet, HCG 92, Pegasus
Seyfert's Sextet, HGC 79, Serpens

NGC 4038 and NGC 4039 are two colliding galaxies
in the constellation Corvus.

UNIVERSAL GRAVITATION **p.201** STAR FORMATION **p.209** DENSITY WAVES AND GALAXY STRUCTURE **p.211**
GALAXY EVOLUTION **p.212**

Starburst Galaxies

GALAXIES BRIGHTENED BY IMMENSE WAVES OF STAR BIRTH

While most irregular galaxies are rich in star-forming gas and dust (as well as newborn stars), some take this to extremes. These so-called starburst galaxies are undergoing bursts of star formation at a vastly accelerated rate. In many cases, starbursts seem to be linked to encounters with larger nearby galaxies – the gravity of the bigger galaxy raises tidal forces that generate a huge rolling wave of compression in the smaller one's gas clouds, triggering the birth of large numbers of stars.

Because the brightest and most massive stars age and die within just a few million years of their formation, the starburst process can sustain itself for long periods. Even as the tidal forces that were its initial cause ease, a chain reaction effect takes over as the supernova shockwaves produced as these massive stars explode enrich and compress nearby gas, triggering new waves of starbirth.

However, supernovae may ultimately also bring an end to the process – they can heat up and accelerate the motion of nearby gas so that it is less likely to be caught by gravity and incorporated into new stars. In extreme cases, this hot gas may overcome the relatively weak gravity of its galaxy entirely and escape into surrounding intergalactic space.

Messier 82 is a starburst galaxy in the constellation Ursa Major about 12 million light years from Earth.

BOUND SYSTEMS **p.18**   PRIMEVAL GALAXIES **p.45**   MERGING GALAXIES **p.46**

Notable starburst galaxies

Silver Coin Galaxy NGC 253, Sculptor

NGC 1569, Camelopardalis

Arp 220, Serpens

NGC 7714, Pisces

ALAN SOLINGER

The concept of starburst galaxies was proposed by Alan Solinger (1941–2017) and colleagues, after a study of the curious galaxy M82 disproved theories that it had been the scene of a vast explosion, and showed that the source of its brightness was different from so-called active galactic nuclei. Instead, they attributed M82's unusual appearance to a stellar baby boom.

Radio Galaxies

GALAXIES EMBEDDED WITHIN LOBES OF RADIO-EMITTING GAS

BOUND SYSTEMS **p.18** PRIMEVAL GALAXIES **p.45** MERGING GALAXIES **p.46**

The giant lenticular radio galaxy NGC 1316 in the Fornax constellation is the fourth brightest radio source in the sky.

Many peculiar galaxies are found at the centre of vast clouds of gas that emit radio waves. Such clouds often take the form of two 'lobes' to either side of the central galaxy, which may be far larger than the galaxy itself. A big clue to the origin of these radio lobes lies in the thin, narrow jets that sometimes link them to the very centre of their parent galaxy.

Most, if not all, large galaxies have a substantial supermassive black hole at their very centre – a huge object with the mass of millions or even billions of Suns that normally remains undetectable except through its gravitational influence. Given the billions of years of history since these galaxies formed, most stars and other material in vulnerable orbits will have been absorbed into the black hole long ago, leaving behind only matter that orbits at a safe distance. Galaxy interactions and close encounters, however, can send material falling towards the black hole, where strong tidal forces tear it apart. A combination of gravity and the black hole's powerful magnetic field can accelerate some of the infalling material and eject it at close to the speed of light in two jets aligned with the black hole's poles. Where these jets collide with surrounding intergalactic gas, they slow down to form huge billowing clouds that glow in radio wavelengths.

DONALD LYNDEN BELL
Donald Lynden Bell (1935–2018) was the first person to suggest that radio galaxies and other active galaxies are powered by a supermassive black hole 'engine'. The proposal built on his earlier model of galaxy formation out of collapsing gas clouds, which suggested the process would frequently produce a monster black hole at the galaxy's centre.

Notable radio galaxies
NGC 5128, Centaurus
Messier 87, Virgo
3C 273, Virgo
NGC 1316, Fornax
3C 48, Hercules

UNIVERSAL GRAVITATION **p.201** DENSITY WAVES AND GALAXY STRUCTURE **p.211** GALAXY EVOLUTION **p.212** ACTIVE GALACTIC NUCLEI **p.213**

Seyferts and LINER Galaxies

GALAXIES WITH UNUSUALLY BRIGHT AND UNPREDICTABLE CORES

Notable Seyfert galaxies
NGC 5128, Centaurus
Messier 77, Cetus
NGC 1097, Fornax
Messier 94, Canes Venatici
ESO 97-G13, Circinus

CARL SEYFERT

Carl Seyfert (1911–60) was the first person to realize that the excess light from spiral galaxies with unusually bright cores is emitted at just a few wavelengths, appearing as narrow emission lines in the galaxy's spectrum with broad 'wings' created as the hot material moves at thousands of kilometres per hour around the core.

When material falls onto a galaxy's central black hole to create an active galactic nucleus (AGN), it can give rise to a variety of effects, alongside or independent of radio emission. As gas and dust fall towards the black hole, the rapid increase in gravity can tear it apart and heat it up, creating a solar-system-sized disc of glowing material that shines out from the centre of the galaxy. As the amount of inflating material varies, so too do the intensity and wavelength of radiation from the galaxy's centre. From a distance, this can give some galaxies, known as Seyfert galaxies, a bright nucleus with a compact, starlike appearance.

Other active galaxies have less distinctive nuclei, but show unusually intense features at specific wavelengths and colours in their spectrum, corresponding to the emission of light by specific elements. These Low-Ionization Nuclear Emission-line Regions (LINERs) may be linked to accelerated rates of star formation around an active nucleus, rather than to the AGN itself.

The intense brightness of M106 in the Canes Venatici constellation makes it a characteristic Seyfert galaxy.

BOUND SYSTEMS **p.18** PRIMEVAL GALAXIES **p.45** MERGING GALAXIES **p.46**

Quasars and Blazars

BLAZING INTERGALACTIC BEACONS POWERED BY VORACIOUS BLACK HOLES

Notable quasars
3C 273, Virgo
3C 48, Triangulum
The Einstein Cross, Pegasus
J0313–1806, Eridanus

MAARTEN SCHMIDT
The nature of quasars was exposed by Maarten Schmidt (b.1929) after he identified a visual counterpart to a mysterious radio source known as 3C 273 and captured its spectrum. This revealed spectral lines that did not match with any known element, but Schmidt soon realized they were actually familiar hydrogen lines, Doppler-shifted out of place as 3C 273 retreats at high speed.

The most spectacular active galaxies, quasars are distant objects that are typically billions of light years away from Earth. In general, they are young galaxies with numerous stars and large amounts of gas and dust being consumed by a still-growing central black hole. Infalling material at the heart of the galaxy is heated to millions of degrees Celsius, emitting radiation that ranges from radio to high-energy X-rays. Light from the AGN swamps that of the host galaxy, giving these objects a star-like appearance through all but the most powerful telescopes – hence the name quasar (short for 'quasi-stellar radio source').

Quasars can take a variety of forms – the luminous central disc is surrounded by a broad doughnut of opaque gas, dust and stars, so if the host galaxy lies edge-on to Earth, the AGN is hidden and only the radio lobes, if any, can be seen. Occasionally, the galaxy presents itself face-on, with jets of particles escaping from the AGN aligned directly towards Earth, giving rise to a blazar, an object with unusual features in its spectrum.

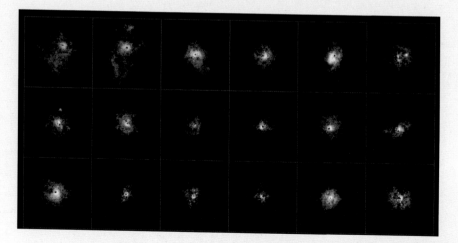

Gas haloes around distant quasars, as viewed from the ESO Very Large Telescope in Chile.

UNIVERSAL GRAVITATION **p.201** DENSITY WAVES AND GALAXY STRUCTURE **p.211** GALAXY EVOLUTION **p.212** ACTIVE GALACTIC NUCLEI **p.213**

The Local Group

THE FAMILY OF GALAXIES LARGE AND SMALL, NEIGHBOURING THE MILKY WAY

The Local Group is the Milky Way's small home galaxy cluster. It contains around 80 known galaxies, most of which are small dwarf systems. Two large spirals – the Milky Way and Andromeda galaxies – dominate the group, with a third smaller spiral (the Triangulum Galaxy) and two mid-sized irregulars (the Magellanic Clouds) as the other significant members. The group forms an outlying and relatively loose 'clump' within the larger Laniakea Supercluster, which is itself centred on the dense Virgo Cluster some 60 million light years from Earth.

The entire system fills a volume of space about 10 million light years across, with most of the smaller galaxies in direct orbit around one of the two major spirals, giving the group an overall dumb-bell-like shape. However, at the edges it blurs into neighbouring clusters and groups, leading to debates about which galaxies are truly Local Group members: technically, a galaxy may fall within the boundaries of the group, but if its orbit suggests it is not gravitationally trapped in orbit, it is not a member.

Meanwhile, the gravitational attraction between the Andromeda and Milky Way spirals is pulling the two galaxies towards each other at a speed of around 179 kilometres per second (400,000 miles per hour), dooming them to an eventual collision in about 4 billion years.

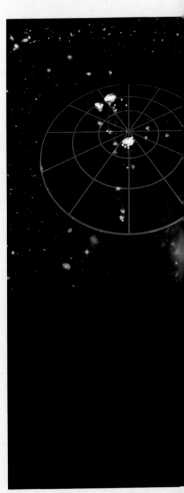

An artist's impression of the solar system and other significant members of the Local Group.

BEGINNINGS OF STRUCTURE **p.40** PRIMEVAL GALAXIES **p.45** MERGING GALAXIES **p.46**
BIRTH OF THE MILKY WAY **p.47**

Notable Local Group members

Milky Way galaxy

Andromeda Galaxy Messier 31

Triangulum Galaxy Messier 33

Large Magellanic Cloud,
Dorado/Mensa

Small Magellanic Cloud, Tucana

Messier 32, Andromeda

Messier 110, Andromeda

GÉRARD DE VAUCOULEURS

The most widely used system for classifying galaxies was developed by Gérard de Vaucouleurs (1918–95), building on the earlier ideas of Edwin Hubble. De Vaucouleurs and his wife Antoinette compiled a detailed catalogue of bright galaxies that took account of spiral elements such as ring- and lens-shaped structures, as well as the size of the galactic core and the coherence of the spiral arms.

UNIVERSAL GRAVITATION **p.201** DENSITY WAVES AND GALAXY STRUCTURE **p.211** GALAXY EVOLUTION **p.212**

The Milky Way

OUR HOME GALAXY – A COSMIC PINWHEEL OF GAS, DUST AND STARS

The Milky Way as it appeared above Fish River Canyon, Namibia, on a night in 2017.

Notable Milky Way satellites
Large Magellanic Cloud, Dorado/Mensa
Small Magellanic Cloud, Tucana
SagDEG (dwarf elliptical/spheroidal), Sagittarius
Canis Major dwarf (irregular), Canis Major
Antlia 2 (star cloud), Antlia

Our home galaxy, the Milky Way is a vast barred spiral system with a visible diameter of at least 120,000 light years, containing between 100 billion and 400 billion stars (the vagueness arises from uncertainties about the extent of hard-to-see outline regions and the number of faint red dwarf stars). Our Solar System lies on a minor 'spur' of one of the spiral arms, about 27,000 light years from the centre which it orbits every 240 million years. From this point of view, stars are far more abundant when we look in directions across the plane of the galaxy than when we look up or down, past stars in our local neighbourhood towards intergalactic space. Hence the Milky Way appears as a pale band of star clouds circling the sky, brightest in the direction of the crowded core.

Because nearby stars, gas and dust block our view at visible wavelengths, astronomers must use other wavelengths to map the Milky Way's more distant areas – and as a result, they are still discovering its true shape. The bar of stars crossing the 20,000 light-year-wide bulge of the core has only been recognized in the past two decades, and there are still disagreements about the number of spiral arms and their relative prominence.

JACOBUS KAPTEYN
Evidence for the rotation of the Milky Way was first gathered by Jacobus Kapteyn (1851–1922) during exhaustive surveys of stars in different parts of the sky. Kapteyn identified two 'streams' of stars, apparently moving in opposite directions. Later astronomers showed this was an illusion caused by stars nearer the galactic centre orbiting more rapidly than our Solar System, while those further out orbit more slowly.

UNIVERSAL GRAVITATION **p.201** STAR FORMATION **p.209** DENSITY WAVES AND GALAXY STRUCTURE **p.211** GALAXY EVOLUTION **p.212**

The Halo and Globular Star Clusters

THE SEEMINGLY EMPTY REGION AROUND THE MILKY WAY, AND THE ENORMOUS BALLS OF STARS THAT ORBIT THROUGH IT

SOLON IRVING BAILEY
Much of the pioneering work on globular clusters was done by Solon Irving Bailey (1854–1931) while working at Harvard Observatory's outpost in Peru. Bailey's meticulous photographic plates revealed many thousands of individual stars within the clusters and ultimately led to the recognition that they are distinctly different from stars in the body of the Milky Way.

The luminous hub and disc of our galaxy are surrounded by a vast, roughly spherical region named the halo, several hundred thousand light years in diameter. While this appears to be mostly empty space, the motions of stars in the outer part of the disc can be explained only by the presence of substantial unseen dark matter whose gravity helps to shape their orbits.

Among this dark matter orbit spherical clouds of stars known as globular clusters. Typically containing hundreds of thousands of stars in a volume perhaps 100 light years across, these clusters are dominated by red and yellow stars, packed so tightly together that in their cores they are separated by just tens of billions of kilometres (mere light days). In these conditions, stellar close encounters, collisions and even mergers are relatively common. The absence of heavy elements in these stars is a sign that these are some of the most ancient stars in the cosmos.

Around 150 globular clusters follow their own long orbits around the Milky Way, nearly all within 100,000 light years of the centre. They are thought to be the relics of vast bursts of star formation as smaller galaxies collided and coalesced to create our galaxy – giant elliptical galaxies, created by an even longer series of mergers and collisions, have many thousands of similar clusters.

Notable globular clusters
Omega Centauri (NGC 5139), Centaurus
47 Tucanae (NGC 104), Tucana
Messier 22, Sagittarius
Messier 5, Serpens
Messier 13, Hercules
Messier 15, Pegasus

NON-LUMINOUS OBJECTS **p.24** BIRTH OF THE MILKY WAY **p.47** COSMIC RECYCLING **p.48**

Easily visible from Earth with binoculars, Messier 4 is a globular
cluster in the constellation of Scorpius.

UNIVERSAL GRAVITATION **p.201** GALAXY EVOLUTION **p.212** DARK MATTER **p.214**

Spiral Arms

**CONCENTRATED ZONES OF STAR FORMATION WINDING ACROSS
THE DISC OF THE MILKY WAY**

EWEN AND PURCELL

Confirmation of the spiral arms came thanks to a breakthrough by physicists Harold Ewen (1922–2015) and Edward Purcell (1912–97). In 1951 they detected a predicted emission at radio wavelengths from neutral hydrogen atoms in space. Although the emission, known as the 21-cm line, is rare, the vast amounts of hydrogen in the Milky Way generate signals strong enough to reveal its structure.

Milky Way spiral arms
Perseus Arm
Scutum-Crux-Centaurus Arm
Carina-Sagittarius Arm
Norma arm
Orion-Cygnus spur

The arms that define the Milky Way's spiral structure are regions where star formation is an ongoing process, or has recently ceased. Although the precise number of arms within our own galaxy and their relative importance are hard to measure from our location within the galactic plane, there are at least two major and two minor arms.

Stars, gas and dust move in and out of these spiral 'traffic jam' zones as they follow their own elliptical orbits around the centre of the galaxy. As they slow down and jostle together, denser regions begin to collapse and develop into star-forming nebulae (along what appears to be the arm's 'trailing edge'), becoming visible as emission nebulae whose gas is excited by radiation from the new-born stars within them. New star clusters dominated by bright but short-lived stars emerge along the leading edge, slowly disintegrating over tens of millions of years as their more sedate and long-lived members join the general disc population. The brightest stars of all, however, age and die before their orbits can carry them out into the wider disc.

Galaxy M51 in the Canes Venatici constellation has a clearly discernible spiral structure.

DIFFUSE MATTER **p.20** STARS **p.22** BIRTH OF THE MILKY WAY **p.47** COSMIC RECYCLING **p.48**

The Galactic Centre

THE HEART OF THE MILKY WAY, WHERE STARS CROWD AROUND A MONSTER BLACK HOLE

ANDREA GHEZ
Andrea Ghez (b.1965) played a leading role in the discovery of the Milky Way's central supermassive black hole. Using infrared imaging to pierce the intervening dust, she tracked the motion of stars orbiting an unseen object at the galactic centre, eventually showing that this object is so massive and compact that it can only be a black hole.

Galactic centre features
Sagittarius A* black hole
S-star cluster
Arches star cluster
Quintuplet star cluster

The inner regions of the Milky Way consist of a vast bulge, roughly 20,000 light years in diameter and 8,000 light years thick, dominated by old, red and yellow stars. While this bulge is mostly quiet and free of star-forming activity, observations in wavelengths such as X-rays and ultraviolet reveal a different story at the very heart of our galaxy, about 27,000 light years away.

The galactic centre is surrounded by several clusters of recently formed stars, including some of the most massive known, swathed in clouds of superheated gas. The turbulence of this region seems to be driven by the gravity of a monstrous black hole with the mass of about 4 million suns. While this black hole would have powered violent quasar-like activity early in the Milky Way's history, it has now mostly cleared its surroundings, leaving stars orbiting at a safe distance. Stray gas falling onto the black hole emits radiation that turns the precise galactic centre into a radio source known as Sagittarius A*.

The relatively subdued lighting around Sagittarius A* is caused by the predominance of superheated gas clouds.

UNIVERSAL GRAVITATION **p.201** STELLAR EVOLUTION **p.207** GALAXY EVOLUTION **p.212** ACTIVE GALACTIC NUCLEI **p.213**

Star-forming Nebulae

VAST CLOUDS OF DARK DUST AND GLOWING GAS FROM WHICH STARS ARE BORN

Stars are born from the collapse of vast clouds of interstellar gas and dust dominated by hydrogen. This 'interstellar medium' has an average density of about 1 atom per cubic centimetre across the Milky Way, but clumps together at a million times this density in clouds known as nebulae. Most nebulae are luminous only in low-energy radio infrared and radio wavelengths, and almost transparent to visible light. However, when they undergo internal collapse due to their own gravity, or external compression (for instance when nebulae collide or nearby stars explode releasing shockwaves), they can become opaque, dust-rich 'dark nebulae'.

Within these nebulae, concentrations of matter begin to pull on their surroundings in a process of collapse that gives birth to stars. Pressure from the fierce radiation of newborn stars and their strong stellar winds (streams of particles blowing off their surfaces) gradually clears the nebulosity from their immediate surroundings until they emerge into visibility. Meanwhile, high-energy ultraviolet radiation from the heaviest and hottest of these newborn stars stimulates the surrounding gas, boosting the internal energy of atoms which then fall back to a less energetic state by emitting visible light. The result is a glowing emission nebula, often with a cavern-like appearance and a brilliant cluster of newborn stars at its centre.

WILLIAM HUGGINS

Confirmation that many nebulae are gaseous came through the work of William Huggins (1824–1910). By splitting their light to form a spectrum, Huggins discovered that some nebulae (now known to be distant galaxies) produced light of nearly every wavelength and thus were probably composed of countless stars, while others emitted light of a few distinct colours, similar to laboratory gas samples.

Star-forming activity in the Tarantula Nebula is so intense that its luminosity belies its vast distance from Earth – 160,000 light years.

DIFFUSE MATTER **p.20** STARS **p.22** COSMIC RECYCLING **p.48**

Notable star-forming regions
Orion Nebula Messier 42, Orion
Carina Nebula NGC 3372, Carina
Lagoon Nebula, Messier 8, Sagittarius
Eagle Nebula, Messier 16, Serpens
Tarantula Nebula, Large Magellanic Cloud, Dorado

UNIVERSAL GRAVITATION **p.201** STELLAR SPECTROSCOPY **p.202** STAR FORMATION **p.209** DENSITY WAVES AND GALAXY STRUCTURE **p.211**

Bok Globules

OPAQUE CLOUDS OF GAS AND DUST THAT GIVE RISE TO INDIVIDUAL STELLAR SYSTEMS

Notable Bok globules
Thackeray's Globules in IC 2944, Centaurus
Barnard 68, Ophiuchus
Pillars of Creation, Messier 16, Serpens
NGC 281, Cassiopeia

BART BOK
Bok globules are named after Bart Bok (1906–83) who, alongside Edith Reilly, first reported their presence in a variety of nebulae in 1947. At the time, Bok predicted that the globules might be the equivalent of stellar cocoons, but it was not until 1990 that infrared observations confirmed the presence of young stars hidden within them.

As star-forming nebulae collapse under the influence of gravity and the pressure of stellar winds from earlier generations of stars, regions of high density go through a period of snowballing growth, until eventually they begin to emerge as distinct globules – clouds of opaque dust and gas about a light year across, known as Bok globules. The gravity of these globules is strong enough to keep a hold on their material even as it is blown away from their surroundings by the fierce radiation and stellar winds emanating from other newly formed nearby stars.

Individual globules can remain attached for some time to larger pillar-like clouds by elongated tendrils – remnants of opaque interstellar matter that lie in the globule's 'shadow' and are therefore protected from the stripping effect of the nearby stars. Eventually, however, these stellar umbilical cords fray and disappear, leaving the globules as isolated clouds in space. Inside them, matter continues to concentrate in one or more dense pockets, eventually giving rise to either a single star or a multiple star system.

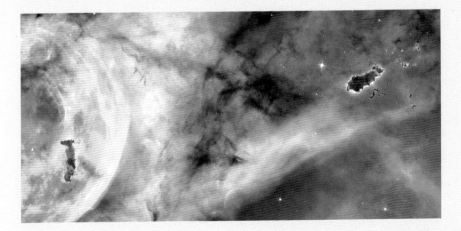

A Bok globule nicknamed 'the caterpillar' can be seen on the right of this image inside the Carina Nebula.

DIFFUSE MATTER **p.20** STARS **p.22** NON-LUMINOUS OBJECTS **p.24** COSMIC RECYCLING **p.48**

Young Stars

UNSTABLE STARS THAT ARE JUST BEGINNING TO SHINE

JOHN RUSSELL HIND
The most famous young star in the sky, designated T Tauri, was discovered by John Russell Hind (1823–95). It was found alongside a small cloud of glowing gas, and both star and nebula faded over the following years. T Tauri's unpredictable changes in brightness are thought to be caused by ejected gas clouds partially blocking its light.

Notable young stars
Trapezium cluster in Messier 42, Orion
T Tauri, HD 284419, Taurus
Gliese 674, Ara
AU Microscopii, HD 197481, Microscopium

New stars are born when the cores of collapsing Bok globules grow hot and dense enough for nuclear fusion to take hold in their cores. Up until this point, the young 'protostar' releases energy purely from gravitational collapse, shrinking to about the size of the planet Jupiter. As its mass becomes concentrated, its rotation speeds up until it is completing each revolution in a matter of hours. This high speed causes it to throw off much of the material still being drawn in by gravity. Ejected matter is channelled into twin jets along the star's axis of rotation, which collide with surrounding interstellar matter to create glowing lobes named Herbig-Haro objects.

Meanwhile, matter continues to concentrate in the core until it is hot and dense enough for nuclear fusion to begin. At first, the process is only viable with deuterium (a form of hydrogen), but as the star's interior grows hotter, normal hydrogen fusion takes over. The outward pressure from radiation causes the star's interior to inflate rapidly, creating an unstable adolescent star with a strong stellar wind of particles blowing off its surface.

The Serpens Nebula, which can be found in Serpens Cauda, is 1,300 light years away. Young stars give this star-forming region its sheen.

COLLISIONAL ACCRETION (SOLAR SYSTEM FORMATION) **p.199** UNIVERSAL GRAVITATION **p.201**
THE MASS–LUMINOSITY RELATION **p.205** THE POWER SOURCE OF STARS **p.206**

Exoplanets

THE MENAGERIE OF PLANETS GREAT AND SMALL NOW KNOWN TO ORBIT OTHER STARS

ORIGINS OF THE SOLAR SYSTEM **p.49** CREATION OF THE PLANETS **p.51**

Artist's impression of the surface of an exoplanet orbiting the red dwarf star Proxima Centauri.

The process of star formation naturally leads to a flattened disc of excess material in orbit around most newborn stars, and in many cases the dust and gas in this disc will clump together to form planets. Most stars in the Milky Way probably have one or more planets in orbit around them, and these exoplanets vary more wildly than the worlds of our own Solar System. The largest exoplanets may have several times the mass of Jupiter (though they are not significantly larger), while in between these gassy worlds and the rocky, Earth-like planets lie a range of intermediate bodies including so-called gas dwarfs ('mini-Neptunes') and solid 'super-Earths'. Conditions on these alien worlds are governed by a mix of their composition and orbit – for example, many of the first exoplanets to be discovered are 'hot Jupiters'. These are gas giants in orbits much closer to their star than Mercury is to the Sun – so close that their atmosphere may be boiling away into space or may even have disappeared completely, leaving an exposed solid core known as a Chthonian planet. While the planets of our Solar System follow near-circular paths around the Sun, many exoplanet orbits are highly elliptical. Planets have been discovered orbiting around binary star pairs, and around the individual stars in multiple systems.

MICHEL MAYOR

The first exoplanet around a Sun-like star was discovered in 1995 by Michel Mayor (b.1942) and Didier Queloz (b.1966). Using ELODIE, an advanced spectroscope, to spread out the colours of starlight, they detected tiny Doppler shifts in the light of the star 51 Pegasi, caused as an orbiting 'hot Jupiter' planet tugs it in different directions.

Notable exoplanets
51 Pegasi b (aka Dimidium), Pegasus
Kepler-1649c, Cygnus
TOI-849b, Sculptor
WASP-69b, Aquarius
Fomalhaut b, Piscis Austereness

COLLISIONAL ACCRETION (SOLAR SYSTEM FORMATION) **p.199** UNIVERSAL GRAVITATION **p.201**
PLANETARY MIGRATION **p.200**

Open Star Clusters

GROUPS OF YOUNG STARS WITH A SHARED ORIGIN, DOMINATED BY BRILLIANT, SHORT-LIVED STARS

The formation of an individual star can be surprisingly rapid compared to the overall scale of stellar lifetimes, and as a result star-forming nebulae behave like factories, mass-producing stars in waves of activity. These stars emerge in groups known as open clusters that are loosely bound together by gravity (in contrast to the more compact and tightly bound globular clusters).

Open clusters typically contain stars with a broad range of masses, colours and luminosities, but the younger a cluster is, the more it will be visually dominated by brilliant blue and white heavyweight stars. These shine far more brilliantly than their lower-mass siblings, but race from birth to death in just a few million years. As the cluster ages, the heaviest stars rapidly come to the end of their lives, transforming into supergiants before destroying themselves completely in supernova explosions. Thus over time only the fainter, cooler and less massive stars remain. Close encounters between stars in the cluster will also gradually disrupt their orbits, leading the cluster to gradually disintegrate, while in extreme cases, encounters involving multiple star systems can fling individual stars into exile as high-speed stellar runaways. For a time, the remnants of a cluster may still be traced across space by tracking the paths of a scattered 'moving group' of stars back to a common origin.

The young open cluster NGC 602 is 5 million years old and resides in the Small Magellanic Cloud, a satellite dwarf galaxy of the Milky Way.

Notable open star clusters

The Pleiades (Messier 45), Taurus
The Hyades, Taurus
Melotte 111, Coma Berenices
The Wild Duck (Messier 11), Scutum
The Beehive (Messier 44), Cancer
The Jewel Box (NGC 4755), Crux

BOUND SYSTEMS **p.18** STARS **p.22** COSMIC RECYCLING **p.48**

RICHARD PROCTOR
Popular astronomy writer Richard Proctor (1837–88) made several
important contributions to our understanding of the heavens,
including the discovery of the 'Ursa Major Moving Group'. This group
of stars, some 80 light years from Earth, has a shared general
motion across the sky, and marks the disintegrating remnant of
a 300-million-year-old open cluster.

UNIVERSAL GRAVITATION **p.201** REDSHIFT AND THE DOPPLER EFFECT **p.203**
THE POWER SOURCE OF STARS **p.206** STAR FORMATION **p.209**

Main-sequence Stars

STARS IN THE MIDDLE PHASE OF THEIR EVOLUTION, WHERE MASS, COLOUR AND LUMINOSITY ARE INTIMATELY LINKED

The blue tinge displayed by the Pleiades, a star cluster in the constellation Taurus, is caused by great heat intensity.

STARS **p.22** COSMIC RECYCLING **p.48** THE SUN IGNITES **p.50** THE DEVELOPMENT OF THE SUN **p.55**

The life expectancy of stars varies hugely depending on their mass, from hundreds of billions of years for the most sedate and low-mass objects to just a few million years for fiercely bright stars with many times the mass of the Sun. However, all stars spend the vast majority of their lives generating light and energy by fusing hydrogen nuclei into helium in their cores. This process marks a long phase of stability in a star's outward characteristics, known as its 'main sequence' lifetime.

During this period, all stars obey a relationship that links their intrinsic brightness (luminosity) to their colour and surface temperature: the more luminous a star is, the hotter its surface and the bluer its colour. Measurements of stellar masses have revealed that a star's main-sequence properties are also linked to its mass: lightweight stars tend to be cool, red and very faint, while heavyweights are hot, blue and very bright. These patterns are due to the way that a star's gravity and core conditions (determined by its mass) influence the rate at which fusion reactions take place: small differences in mass create large differences in energy output, and while more massive stars and luminous stars are physically larger than their low-mass cousins, their surfaces are still heated to far higher temperatures by escaping energy.

EJNAR HERTZSPRUNG

The main sequence relationship was first identified by chemist and self-taught astronomer Ejnar Hertzsprung (1873–1967). In 1911, he published a diagram comparing the spectral type (colour) and apparent brightness of stars in the Pleiades cluster. By assuming they all lay at the same distance, he demonstrated that hotter and bluer cluster stars were intrinsically more luminous.

Brightest main-sequence stars
Sirius, Canis Major
Rigil Kentaurus, Centaurus
Vega, Lyra
Procyon, Canis Minor
Achernar, Eridanus
Altair, Aquila

STELLAR SPECTROSCOPY **p.202** STELLAR STRUCTURE **p.204** THE MASS–LUMINOSITY RELATION **p.205** THE POWER SOURCE OF STARS **p.206**

Red and Brown Dwarfs

THE FAINTEST, MOST ABUNDANT AND LONGEST-LIVED STARS

WILLEM JACOB LUYTEN
Many nearby red dwarfs were discovered by Willem Jacob Luyten (1899–1994). He used photographic plates to measure the rate at which stars drift across the sky and worked on the assumption that stars showing larger apparent motions would generally be closer to Earth, allowing him to distinguish luminous but distant stars from faint nearby ones.

Stars with less than half the Sun's mass behave very differently from their heavier siblings. Their cores burn hydrogen at a much slower rate, and hence they appear very faint and have much cooler, redder surfaces. These red dwarfs vastly outnumber all the galaxy's more luminous stars. The closest star beyond our own Solar System, Proxima Centauri, is a red dwarf that can only be seen through substantial telescopes despite lying just 4.25 light years away.

In contrast to other stars, the internal structure of red dwarfs allows their cores to be replenished from their upper layers. This means that, despite containing less material than other stars, they have a larger fuel supply and can theoretically shine for hundreds of billions of years.

Normal nuclear fusion (using the simplest and most abundant form of hydrogen) is only possible within stars that have more than 8 per cent of the Sun's mass, so this is as small as a red dwarf can get. Below this mass lie brown dwarfs – 'failed stars' that may shine weakly by fusing deuterium, a scarce form of hydrogen.

Despite their low energy output, red dwarfs can be surprisingly violent – their churning layers generate tangled magnetic fields far more intense than those on the Sun and larger main-sequence stars. These fields can release intense stellar flares that briefly outshine the star's entire normal energy output.

BOUND SYSTEMS **p.18** STARS **p.22** NON-LUMINOUS OBJECTS **p.24**

Nearest red dwarfs
Proxima Centauri, 4.24 light years, Centaurus
Barnard's Star, 5.96 l.y., Ophiuchus
Wolf 359, 7.86 l.y., Leo
Lalande 21185, 8.31 l.y., Ursa Major
Luyten 726-8 A & B, 8.79 l.y., Cetus

Artist's impression of the 'cool' surface
of a red dwarf star.

 STELLAR STRUCTURE **p.204** THE MASS–LUMINOSITY RELATION **p.205** THE POWER SOURCE OF STARS **p.206**
STELLAR EVOLUTION **p.207**

Monster Stars

HEAVYWEIGHT STARS THAT RACE THROUGH BRIEF, BRILLIANT LIVES

Stars with a mass much greater than the Sun shine far more brightly and tend to have hotter surfaces. The higher temperature and pressure in their cores cause nuclear fusion reactions to run at an accelerated rate, so that stars with tens of times the mass of the Sun can pump out hundreds of thousands of times more energy. This means that, despite beginning their lives with greater amounts of hydrogen in their core, heavyweight stars squander their fuel supplies and may only live for a few million years – perhaps one-thousandth the lifetime of a star like the Sun.

In most stellar heavyweights (with ten or more times the mass of the Sun) the vast amounts of energy being released heat their surfaces to temperatures of 20,000 degrees Celsius or more – so hot that they appear blue and emit much of their radiation as invisible, high-energy ultraviolet rays. At even higher masses, the pressure from radiation is so great that, despite increased energy output, the star's surface may actually be cooler (since it is swollen to such a size that less energy is escaping through each spot on its surface). Such supergiant stars are the brightest in the present Universe, and are found in a variety of colours.

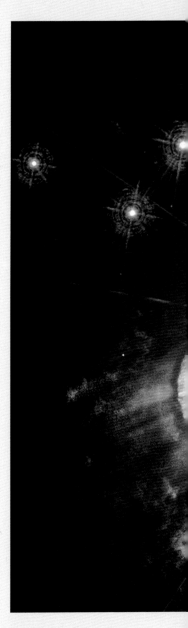

Although 7,500 light years away, in the 1840s the enormous Homunculus Nebula was the second brightest light in the night sky.

MORGAN, KEENAN & KELLMAN

The modern system used for classifying stars was developed in the 1940s and 1950s at Yerkes Observatory, Wisconsin, USA. William Wilson Morgan (1906–94), Philip C. Keenan (1908–2000) and Edith Kellman (1911–2007) improved the system of letters already used for spectral (colour) classification with the addition of Roman numerals to indicate 'luminosity class', ranging from I for the brightest supergiants to V for main-sequence stars.

Brightest massive stars

Canopus, Carina
Rigel, Orion
Hadar, Centaurus
Antares, Scorpius
Deneb, Cygnus

BOUND SYSTEMS **p.18** STARS **p.22** THE FIRST STARS **p.43**

Wolf–Rayet Stars

MASSIVE STARS THAT TEAR THEMSELVES APART THROUGH THEIR INTENSE RADIATION

The intense pressures and searing temperatures at the heart of the most massive stars open a route for more efficient forms of nuclear fusion, allowing stars with masses much higher than the Sun's to shine hundreds of thousands of times more fiercely. In the most extreme stars, the outward pressure exerted by escaping radiation from the core becomes so great that it can overcome the star's gravitational hold on its outer layers. The result is a star with a ferocious 'stellar wind' that sheds material into surrounding space millions of times more rapidly than a Sun-like star. Wolf–Rayet stars – named for their French discoverers, astronomers Charles Wolf (1827–1918) and Georges Rayet (1839–1906) – are a rare class of high-mass star whose outer layers are being cast off into a surrounding nebula while hotter material from the interior is becoming exposed at the surface. Ultimately, they may shed tens of Suns' worth of matter (up to half their mass) over a few million years, with significant effects on the final stages of their evolution.

This battle between the inward pull of gravity and the outward pressure of radiation imposes an upper limit on the mass of stars. Above around 150 solar masses, any star coalescing from the current mix of cosmic raw materials will shine so fiercely that it will blow itself apart before its formation is complete.

JOHN B. HUTCHINGS
The fierce stellar winds around the galaxy's hottest stars were first identified by astronomer John B. Hutchings (b.1941). Hutchings measured dark spectral lines formed in the stellar atmospheres and found their positions were substantially 'blue-shifted' compared to lines generated on the star's surface, because the material that produces them is accelerating towards Earth at hundreds of kilometres per second.

BOUND SYSTEMS **p.18** STARS **p.22** THE FIRST STARS **p.43**

WR 124 is a Wolf–Rayet star in the constellation of Sagitta. One of the fastest runaway stars in the Milky Way, it is surrounded by a nebula of expelled material known as M1-67.

Notable Wolf–Rayet stars
Gamma-2 Velorum (WR 11), Vela
Theta Muscae (WR 48), Musca
WR 22, Carina
R136a1 in the Large Magellanic Cloud, Dorado

STELLAR STRUCTURE **p.204** THE MASS–LUMINOSITY RELATION **p.205** THE POWER SOURCE OF STARS **p.206**
STELLAR EVOLUTION **p.207**

Variable Stars

STARS WHOSE BRIGHTNESS CHANGES IN CYCLES, OFTEN DUE TO INTERNAL INSTABILITY

ARTHUR EDDINGTON
The mechanism underlying most pulsating stars was discovered by Arthur Stanley Eddington (1882–1944), who also created the first accurate models of stellar structure. Eddington realized that a star's core power source must be broadly stable and could not directly cause the most dramatic brightness changes – instead, an outer layer must be acting as a shutter to adjust the amount of light escaping.

While most of the stars in Earth's night sky appear to shine quite steadily, many are not as stable as they first appear – long-term observations or measurements with sensitive instruments reveal that they are prone to varying in brightness. While many of these variations are simply caused by close binary stars passing in front of each other and blocking out each other's light, many more are due to the physical properties of individual stars themselves. For instance, when large dark or bright spots on the surface of a star rotate in or out of view, this can have a significant effect on the amount of light reaching Earth. Similarly, some fast-spinning stars bulge outwards at the equator, giving them an ovoid shape that causes them to vary in brightness depending on how they are seen from Earth.

The most common cause of variability, however, is stellar pulsation – a cyclical change in the amount of light escaping from a star. Variations in luminosity are often accompanied by changes in the size, temperature and even colour of stars, but are unrelated to the energy production from the core itself. Instead, they tend to arise as stars pass through certain life stages where they are prone to developing layers whose opacity can switch off or on depending on their precise temperature.

BOUND SYSTEMS **p.18** STARS **p.22**

Notable variable stars
Mira, pulsating red giant, Cetus
Algol, eclipsing binary system, Perseus
Delta Cephei, pulsating yellow giant, Cepheus
Eta Carinae, unstable supergiant, Carina

Around 6,000 light years from Earth, RS Puppis is one
of the brightest and most variable visible stars.

STELLAR STRUCTURE **p.204**   THE MASS–LUMINOSITY RELATION **p.205**   STELLAR EVOLUTION **p.207**
STELLAR STELLAR NUCLEOSYNTHESIS **p.208**

Binary and Multiple Stars

PAIRS AND GROUPS OF STARS THAT ARE LOCKED IN ORBIT AROUND EACH OTHER

BOUND SYSTEMS **p.18** STARS **p.22**

Around 380 light years away, Albireo is the double star most easily visible from Earth with the naked eye.

E. C. PICKERING

The first detailed calculations of binary orbits were made by Edward Charles Pickering (1846–1919) at Harvard Observatory and Carl Vogel (1841–1907) in Germany. While Pickering discovered the first indivisible 'spectroscopic binary', patchy measurements of its spectrum led him to model its orbit wrongly. Vogel, meanwhile, conducted a more intensive series of observations that led him to the correct solution.

Prominent binary stars

Mizar A/B and Alcor, Ursa Major
Albireo, Cygnus
Epsilon Lyrae, Lyra
Beta Capricorni, Capricornus
Gamma Leporis, Lepus

Many of the Milky Way's brighter stars are members of binary or multiple systems. In most cases, these stellar pairs and larger groups form when a coalescing nebula separates into two or more distinct centres of mass, which accumulate material and ignite into stars while remaining in orbit around each other. On rare occasions, however, crowded conditions inside star-forming nebulae or dense globular clusters can cause stars to be captured into orbit around each other, or even to exchange partners.

Binary systems provide valuable information that reveals the secrets of stellar evolution. Their orbits can disclose their relative masses, and because the stars are at the same distance from Earth we can easily see how this relates to their luminosities. What's more, because in most cases their stars formed at the same time from the same raw materials, binaries offer a neat demonstration of how differences in the mass of stars give rise to differences in their colour and brightness, and even drive their evolution at different rates.

While some stars may take thousands of years to orbit each other, other binary pairs are so tightly bound that they orbit each other in hours or days. These pairs are impossible to split through even the most powerful telescope, but can be detected from clues in the spectrum of their combined starlight.

UNIVERSAL GRAVITATION **p.201** STELLAR SPECTROSCOPY **p.202** REDSHIFT AND THE DOPPLER EFFECT **p.203** THE MASS–LUMINOSITY RELATION **p.205**

Red Giants

**AGEING STARS FAR LARGER AND MORE BRILLIANT THAN THE SUN,
BUT ALSO COOLER AND REDDER**

As a star nears the end of its life, it eventually exhausts the supply of hydrogen available for nuclear fusion in its core. This triggers a number of internal changes to the star – the faltering pressure of radiation pushing out from the centre allows layers above the core to collapse onto it, heat up and begin to fuse themselves. Perhaps surprisingly, this 'shell fusion' actually causes the star to brighten considerably, while the renewed outward pressure causes the star's outer layers to billow outwards. The result is a brilliant star with a significantly cooler surface than it had during its main-sequence lifetime: a red giant. Around 5 billion years from now, this will be the fate of our own Sun, as it swells to engulf the orbits of Mercury, Venus, and perhaps even Earth itself.

Red giants go through several distinct evolutionary phases – in most stars, the core will eventually become hot enough to reignite, fusing the helium left behind by hydrogen fusion into heavier elements such as carbon, nitrogen and oxygen. Once the fuel supply in the core runs out, helium fusion will in turn move out into a shell. The overall result is a star caught in a delicate internal balancing act that lends itself to instability, with long periods of pulsation in both size and brightness.

ERNST ÖPIK

The secrets of red giants were discovered by Ernst Öpik (1893–1985), when he made the revolutionary suggestion that most stars are not 'well mixed'. In other words, their cores have a limited fuel supply that cannot be replenished from elsewhere. This limits stellar lifetimes, but also means the core grows hotter with age, triggering changes that ultimately generate fusion shells.

BOUND SYSTEMS **p.18** STARS **p.22** DEATH OF THE SUN **p.57**

Brightest red giants and supergiants
Arcturus, Boötes
Capella, Auriga
Betelgeuse, Orion
Aldebaran, Taurus
Antares, Scorpius

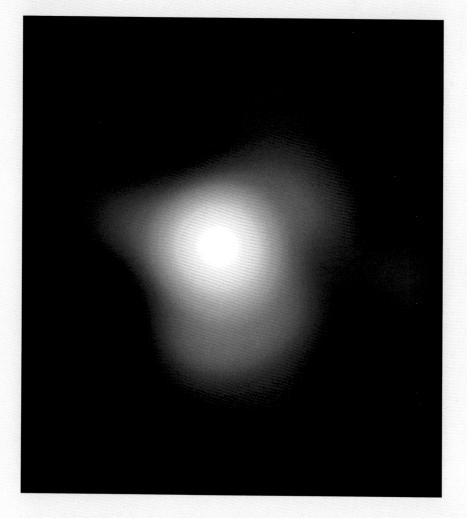

In the Orion constellation 724 light
years from Earth, the red supergiant
Betelgeuse is almost 1,000 times bigger
than the Sun.

 STELLAR STRUCTURE **p.204** THE MASS–LUMINOSITY RELATION **p.205** STELLAR EVOLUTION **p.207**
STELLAR STELLAR NUCLEOSYNTHESIS **p.208**

Planetary Nebulae

BEAUTIFUL SHELLS OF GLOWING GAS, EJECTED AND ENERGIZED BY DYING STARS

Planetary nebulae mark a short-lived but beautiful final phase in the lives of stars like the Sun. As the expanding fusion shells within a red giant work their way towards the surface, the swollen star becomes increasingly unstable and begins to fling its outer layers into surrounding space. These gas bubbles would soon fade and cool beyond visibility, were it not for intense radiation from the remains of the star at their centre. As the star's inner layers are exposed, its surface temperature rises by tens of thousands of degrees Celsius, so that much of its radiation is emitted in the ultraviolet, energizing the surrounding gas for a few thousand years, so that it glows in a similar way to the emission nebulae out of which stars are born.

The shapes of planetary nebulae vary wildly, from simple rings to double-lobed 'butterflies' and tangled, shell-like structures. These forms are determined not only by the properties of the dying star, but also by its surrounding environment. For instance, the nebula may be pinched in at the centre if the star is surrounded by a layer of thicker gas and dust ejected in earlier times, or formed into overlapping spirals where the gravity of a companion star pulls the central red giant in different directions.

Notable planetary nebulae
Ring Nebula, Messier 57, Lyra
Dumbbell Nebula, Messier 27, Vulpecula
Bug or Butterfly Nebula, NGC 6302, Scorpius
Saturn Nebula, NGC 7009, Aquarius
Cat's Eye Nebula, NGC 6543, Draco

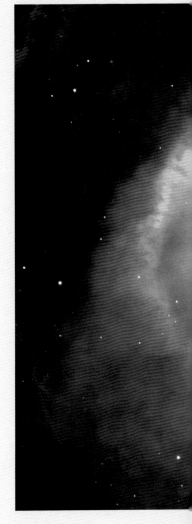

The Helix Nebula (NGC 7293) in Aquarius is formed by a star shedding its outer layers near the end of its evolution. It is sometimes known informally as 'the Eye of God'.

BOUND SYSTEMS **p.18** STARS **p.22** COSMIC RECYCLING **p.48** DEATH OF THE SUN **p.57**

IOSIF SHKLOVSKY

The nature of planetary nebulae was deduced by Soviet astronomer Iosif Shklovsky (1916–85) after measurements revealed their rapid expansion for the first time. Shklovsky realized that they must be very short-lived, and so probably marked a brief transition between two more widespread and long-lived objects. His detective work eventually pinpointed these as red giants and hot, faint white dwarf stars.

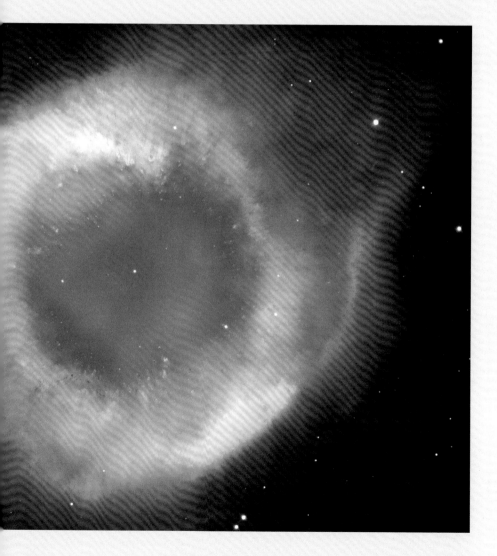

 THE MASS–LUMINOSITY RELATION **p.205** THE POWER SOURCE OF STARS **p.206** STELLAR EVOLUTION **p.207** STELLAR STELLAR NUCLEOSYNTHESIS **p.208**

White Dwarfs

**THE INTENSELY HOT EMBER OF A STELLAR CORE THAT SURVIVES
THE DEATH OF STARS LIKE THE SUN**

Sirius A, the brightest star of all, has been known since antiquity.
Sirius B, on the right, is a white dwarf discovered in 1862.

STARS **p.22** SUBATOMIC PARTICLES **p.28** FUNDAMENTAL FORCES **p.30** DEATH OF THE SUN **p.57**

Nearest white dwarfs
Sirius B, 8.6 light years, Canis Major
Procyon B, 11.4 l.y., Canis Minor
Van Maanen's Star, 14.0 l.y., Pisces
LP 145-141, 15.1 l.y., Musca
40 Eridani B, 16.3 l.y., Eridanus

What happens to stars when they die? Stars of less than eight solar masses meet their end when they cast off their outer layers as a planetary nebula, exposing an exhausted core rich in elements such as helium, carbon, nitrogen and oxygen. With no fusion process left to generate radiation and outward pressure, this core collapses inwards until forces of repulsion between its subatomic particles grow strong enough to halt the process. The resulting core has become a white dwarf – a dense, roughly Earth-sized object that still emits intense radiation as it slowly cools from an initial temperature of perhaps 200,000 degrees Celsius.

The tiny size of white dwarfs makes them inconspicuous – the first to be discovered, such as the famous Sirius B, were spotted due to their gravitational influence over brighter companions in binary systems. In the right circumstances, a white dwarf can even tug gas away from its neighbour, surrounding itself with hot gas that eventually detonates to create a brilliant eruption known as a nova. Some nova systems can produce repeated outbursts, while in other systems, a particularly heavy white dwarf may accumulate so much stolen material that it undergoes a sudden and dramatic collapse into a superdense neutron star, accompanied by a vast explosion known as a Type Ia supernova.

SUBRAHMANYAN CHANDRASEKHAR

The physics of white dwarfs was largely established by Subrahmanyan Chandrasekhar (1910–95). He addressed the long-standing question of how a hot white star could also be small and faint by using new concepts from quantum physics to show how, once internal energy production ceased, a star would collapse until repulsion between electron particles within caused it to stabilize.

STELLAR SPECTROSCOPY **p.202** STELLAR EVOLUTION **p.207** STELLAR NUCLEOSYNTHESIS **p.208**

Supernovae

VAST EXPLOSIONS THAT MARK THE DEATH OF MASSIVE STARS AND FORGE THE UNIVERSE'S HEAVY ELEMENTS

WALTER BAADE
Supernovae are so rare that their existence was only noticed once the distance to other galaxies was established in the 1920s. Walter Baade (1893–1960) and Fritz Zwicky (1898–1974) revisited records of a faint 'nova' seen in the Andromeda Galaxy in 1885, and found that in reality it must have been tens of millions of times brighter than the Sun.

The death of monster stars (with masses greater than eight Suns) is marked by a rare supernova explosion that may briefly outshine an entire galaxy. Dying supergiants with sufficient mass can continue the process of nuclear fusion beyond the limits seen in lesser stars, fusing carbon, nitrogen and oxygen to make elements such as neon, sulfur and iron. Each new wave of fusion generates less energy and exhausts itself more quickly, before moving out into a thin fusion shell around the stellar core. Above this lies a massive outer envelope of hydrogen that becomes increasingly unstable.

A crisis comes when the core attempts to fuse iron – the lightest element for which fusion absorbs, rather than releases, energy. What happens next is complex and still not entirely understood, but the most obvious result is that the radiation pressure supporting the star's structure is abruptly cut off. The core collapses suddenly in on itself, with the overlying layers dropping inwards, compressing and then suddenly rebounding. As the resulting outward shockwave tears the star apart, it compresses and heats the star's upper layers and outer envelope to temperatures far higher than any normal stellar core. The result is a frenzy of nuclear fusion that burns through many solar masses of material in just a few weeks.

Notable galactic supernovae
Supernova of 185 AD, Circinus
Supernova of 1006, Lupus
Crab Supernova, SN 1054, Taurus
Tycho's Supernova, SN 1572, Cassiopeia
Kepler's Supernova, SN 1604, Ophiuchus

BOUND SYSTEMS **p.18** STARS **p.22** SUBATOMIC PARTICLES **p.28** FUNDAMENTAL FORCES **p.30**

Supernova 1987A lies in the Large Magellanic Cloud on the
border of the Dorado and Mensa constellations. Its name refers
to the year of its discovery.

Supernova Remnants

CLOUDS OF SUPERHEATED GAS EXPANDING FROM THE SITES OF SUPERNOVA EXPLOSIONS

As a supernova explosion fades from view, it leaves behind a vast cloud of expanding superheated debris, enriched with the heavy elements that can only form in the energy-rich conditions provided by a supernova. Some of these supernova remnants are visible only through X-rays and other high-energy forms of radiation, but others shine partially in visible light.

At the centre of the expanding cloud, a relic of the star's core survives. With a mass greater than 1.4 Suns, its gravity is strong enough to overcome the pressure that usually stabilizes collapsing stellar cores as planet-sized white dwarfs. As the collapse continues, it is usually halted by even more powerful forces, resulting in a neutron star. The most massive stellar cores of all, however, dwindle to a single point of infinite density, creating a black hole.

Supernova remnants play a key role in the evolution of other stars and solar systems. As they expand and mix with the general interstellar medium they add heavier elements that accelerate the rate of fusion in later generations of stars, and which are also a vital part of rocky planets like Earth. The expanding shockwaves themselves can also act as a trigger for new star formation as they ripple through and compress surrounding nebulae.

FRED HOYLE

The processes that build heavy elements within massive stars and supernovae were first outlined by Fred Hoyle (1915–2001). Hoyle discovered the mechanism by which helium nuclei fuse to form carbon and elements up to iron within supergiant stars, before co-authoring a pivotal 1957 paper that showed how even heavier elements could be assembled within the supernova environment.

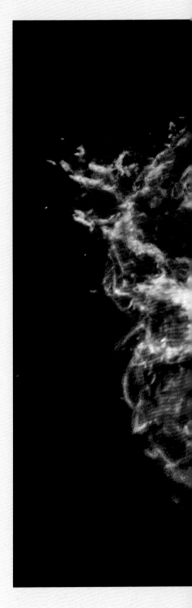

Cassiopeia A is the remnant of a supernova explosion that appeared in our sky 300 years ago.

DIFFUSE MATTER **p.20** ELEMENTS **p.26** SUBATOMIC PARTICLES **p.28** COSMIC RECYCLING **p.48**

Prominent supernova remnants
Crab Nebula, Messier 1, Taurus
Cygnus Loop (Sharpless 103), Cygnus
Tycho's SNR (SN 1572), Cassiopeia
Kepler's SNR (SN 1604), Ophiuchus
Vela SNR (Gum 16), Vela

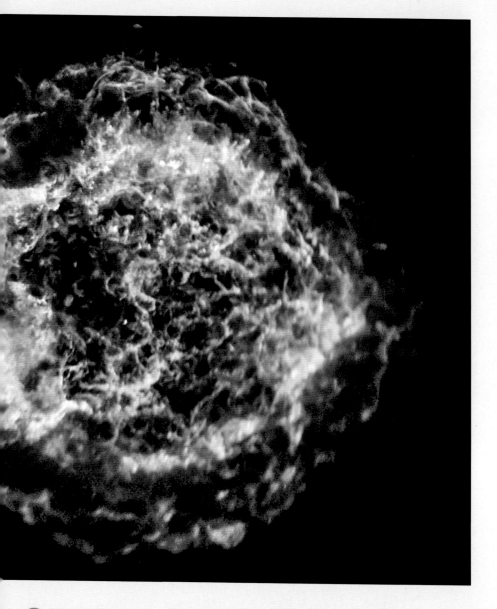

STELLAR EVOLUTION **p.207**   STELLAR NUCLEOSYNTHESIS **p.208**   STAR FORMATION **p.209**
DENSITY WAVES AND GALAXY STRUCTURE **p.211**

Pulsars

SUPERDENSE NEUTRON STARS WHOSE ROTATION SWEEPS BEAMS OF RADIATION ACROSS THE GALAXY

When the core of a monster star collapses during a supernova explosion, the resulting neutron star is one of the Universe's most extreme objects. Compressing over 1.4 Suns-worth of mass into a sphere perhaps 20 kilometres (12.5 miles) across, its material is so dense that a pinhead would weigh as much as a fully laden supertanker.

Nevertheless, certain laws of physics must still be obeyed, and specifically the core's angular momentum (a property related to its original rate of spin and distribution of mass) and magnetic field must be preserved. In practice, this means that many newly formed neutron stars will typically spin very rapidly (hundreds of times per second) and some are wrapped in magnetic fields so intense that they channel most of the radiation escaping from the surface into two narrow beams that emanate from the magnetic poles.

Because the star's axis of rotation and magnetic field are rarely perfectly aligned, the result is a pulsar – an object that sends lighthouse-like beams of radiation sweeping around the heavens, blinking rapidly on and off from the point of view of anyone who happens to be in its path. Despite the intensity of their radiation, the tiny size of neutron stars would render them mostly undetectable if it weren't for these blinking beacons.

JOCELYN BELL BURNELL
The first pulsar was discovered by chance in 1967 while Jocelyn Bell Burnell (b.1943) was attempting to study quasars. Working with her PhD supervisor, she soon realized that precise bursts of radio waves coming from the constellation Vulpecula could be the product of a rapidly rotating neutron star, as predicted in a model published just weeks earlier.

STARS **p.22** SUBATOMIC PARTICLES **p.28** FUNDAMENTAL FORCES **p.30**

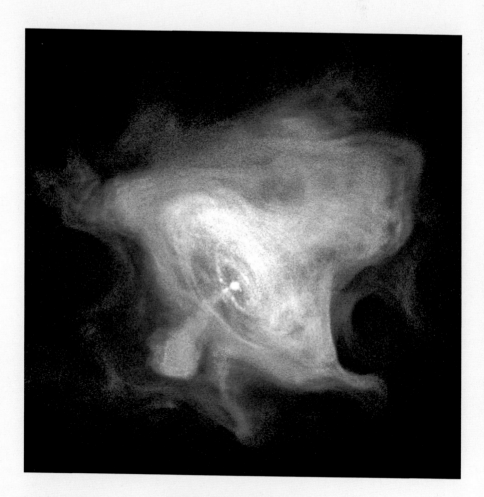

The Crab Pulsar (PSR B0531+21) in the Taurus constellation is a relatively young neutron star. It is a remnant of the supernova SN 1054.

Notable pulsars and their periods
PSR B1919+21 (Bell's pulsar in Vulpecula), 1.337 seconds
PSR B0531+21 (Crab pulsar in Taurus), 33 milliseconds
PSR J0835-4510 (Vela pulsar), 89.33 ms
PSR J1748-2446ad (Fastest known pulsar in Sagittarius), 1.4 ms

UNIVERSAL GRAVITATION **p.201** STELLAR EVOLUTION **p.207**

Stellar Black Holes

COLLAPSED CORES OF MONSTER STARS, WITH GRAVITY SO STRONG THAT NOT EVEN LIGHT CAN ESCAPE

While most supernovae result in the formation of neutron stars, a few generate objects that are even more extreme – black holes. These strange phenomena are formed when the collapsing stellar core is above a certain threshold (8 times the solar mass) so that not even the pressure of neutrons can prevent their collapse. The core's entire mass dwindles to a tiny region of space with infinite density – an object known as a singularity in which the equations of general relativity that describe the normal Universe break down.

The singularity is sealed off from the Universe by a barrier named the event horizon, marking the volume of

Computer simulation of light around a black hole.

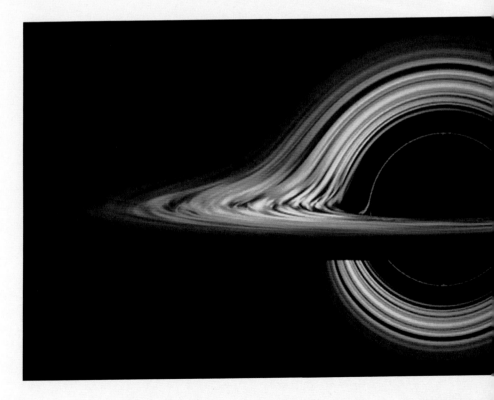

BOUND SYSTEMS **p.18** NON-LUMINOUS OBJECTS **p.24** FUNDAMENTAL FORCES **p.30** THE FATE OF THE UNIVERSE **p.59**

Notable black holes
QV Telescopium, 5 solar masses
V404 Cygni, 9 solar masses
Cygnus X-1, 21.2 solar masses
Sagittarius A*, 4 million solar masses
M87* (Virgo A), 6.5 billion solar masses

space where gravity is so strong that not even light can escape. While a black hole itself is invisible, it can give itself away through its gravitational effect on other objects, such as neighbouring stars. It may also produce tell-tale signals when it pulls in matter that strays too close (or drags it from the outer atmosphere of a neighbour). As material spirals down towards the event horizon, it is torn apart and heated to intense temperatures by the rapidly increasing gravitational field, producing a dying burst of X-rays and other radiation before it is swallowed up and added to the singularity.

KARL SCHWARZSCHILD
The potential existence of black holes was predicted by Karl Schwarzschild (1873–1916) as a possible consequence of Einstein's Theory of General Relativity, published in 1915. However, it was not until 1971 that astronomers Louise Webster, Paul Murdin and Charles Thomas Bolton identified the 'X-ray binary' system Cygnus X-1, now confirmed to contain a stellar-mass black hole.

SPECIAL RELATIVITY **p.196** GENERAL RELATIVITY **p.197** UNIVERSAL GRAVITATION **p.201**
STELLAR EVOLUTION **p.207**

The Solar System

THE SPACE AROUND THE SUN – AND ALL THAT IT CONTAINS

Our Solar System is defined as the region dominated by the influence of our local star, the Sun – but what sort of influence? Traditionally, it's been seen as the family of objects in thrall to the Sun's gravity, heat and light. As well as the Sun itself, this includes eight major planets, the satellites and rings that orbit them, a handful of smaller but still substantial dwarf planets, and countless even smaller objects, including rocky asteroids and icy comets. By this measure, the outer edge of the Solar System extends to the Oort Cloud – a vast spherical halo of dormant comets at the limits of the Sun's gravitational grasp. Most

EUGENE PARKER
The existence of the solar wind filling the Solar System was first proposed by Eugene Parker (b.1927) in 1957, a few years before it was detected by early spacecraft. Parker realized that extreme temperatures in the Sun's upper atmosphere would cause it to shed particles, which could explain effects such as the tails of comets always pointing away from the Sun.

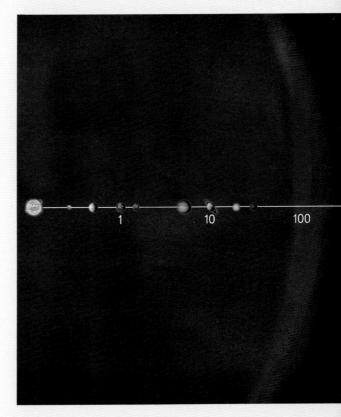

BOUND SYSTEMS **p.18** ORIGINS OF THE SOLAR SYSTEM **p.49** CREATION OF THE PLANETS **p.51**
THE DEVELOPMENT OF THE SUN **p.55**

distances in the Solar System are easily measured in terms of the Astronomical Unit (AU) – the average Earth–Sun distance of about 149.6 million kilometres (93 million miles). In these terms, the Oort Cloud stretches to at least 63,000 AU from the Sun (about a light year).

An alternative definition of the Solar System limits it to the region within which the solar wind (a stream of particles blowing out from the Sun at supersonic speeds) is dominant. This wind envelops the orbits of the planets, but slows down beyond Neptune as it encounters pressure from gas and dust in interstellar space. It comes to a halt in a zone known as the heliopause, 120 AU from the Sun.

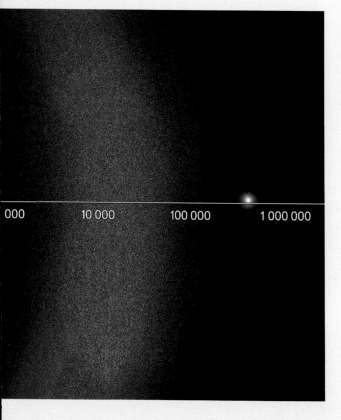

000 10 000 100 000 1 000 000

Limits of the Solar System

Orbit of Neptune: 5 billion kilometres (2.8 billion miles)

Edge of Kuiper Belt: 15 billion kilometres (8.4 billion miles)

Heliopause: 18 billion kilometres (9.6 billion miles) minimum

Edge of Oort Cloud: c.7.5 trillion kilometres (4.5 trillion miles)

The Solar System on a logarithmic scale, with each division beyond 1 AU representing 10 times the previous distance. From the Sun on the left of the image, the planets shown are Mercury, Venus, Earth, Mars, Jupiter, Saturn, Uranus and Neptune.

COLLISIONAL ACCRETION (SOLAR SYSTEM FORMATION) **p.199** UNIVERSAL GRAVITATION **p.201**
THE POWER SOURCE OF STARS **p.206**

The Sun

THE MAIN-SEQUENCE STAR AT THE CENTRE OF THE SOLAR SYSTEM

Layers of the Sun
Core (c.170,000 kilometres/
105,000 miles radius)
Radiative zone (c.480,000
kilometres/298,000 miles)
Convective zone (c.696,000
kilometres/432,500 miles)
Photosphere (c.300 kilometres/
186 miles thick)
Chromosphere
(c.3,000 kilometres/1,864 miles)
Transition region
(c.100 kilometres/62 miles)
Corona (millions of kilometres)

The Sun is our local star – a vast ball of gas with a visible diameter of 1.4 million kilometres (870,000 miles) that contains 99.8 per cent of the Solar System's total mass. It is a fairly average low-mass star, roughly 4.6 billion years old and about halfway through the main-sequence lifetime in which it shines by the fusion of hydrogen into helium. High-energy radiation escaping from the core takes tens of thousands of years to reach the surface through two internal layers known as the radiative zone and the convective zone. In the first, radiation ricochets in a thick fog and gradually loses energy as it makes its way slowly outwards. At the base of the upper layer, radiation is absorbed and heats the gas-like plasma of the Sun's interior, causing it to rise to the surface in vast convection cells. The Sun's visible surface or photosphere marks the layer at which it becomes transparent – radiation is released from the top of the cells and races away into space, while the cooling material sinks down and the cycle repeats. The Sun's sparse outer layers continue well beyond the photosphere, forming a vast outer atmosphere known as the corona, which merges seamlessly into the solar wind blowing out across the Solar System.

DOUGLAS GOUGH
While much of our knowledge of solar structure is theoretical, the existence of its internal layers is backed by helio-seismology, a technique pioneered by Douglas Gough (b.1941). In the 1970s, Gough measured oscillations in the Sun's surface caused by sound waves moving through its interior. He used this to map solar structure in the same way as geologists use seismic waves on Earth.

STARS **p.22** ORIGINS OF THE SOLAR SYSTEM **p.49** THE SUN IGNITES **p.50**
THE DEVELOPMENT OF THE SUN **p.55**

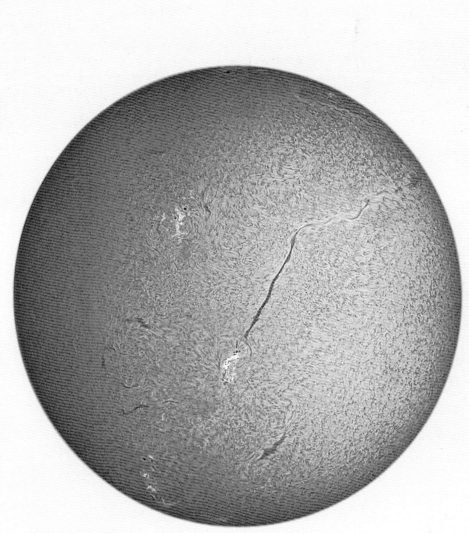

The Sun photographed through a
narrow filter that makes visible the
dark filaments in the area above the
photosphere of the star.

Solar Activity

CYCLICAL CHANGES TO THE SUN'S APPEARANCE AND ACTIVITY

GEORGE ELLERY HALE
The key role of magnetism in driving solar activity was discovered by George Ellery Hale (1868–1938). In 1908 he analysed light from around sunspots, and discovered that it was split into two polarized images of slightly different wavelengths. He attributed this to the Zeeman effect – the action of a strong magnetic field on light.

Although the Sun superficially appears to be a featureless disc, special observing and photographic techniques can reveal a multitude of features on and above its surface. Many of these alter their intensity, frequency and position in the course of a roughly 11-year solar cycle linked to the creation and destruction of magnetic fields in the Sun's outer layers.

Most prominent are the sunspots – dark patches in the photosphere that can approach the size of Earth itself, and which are carried around the disc as the Sun rotates. The spots appear dark because they are cooler than their surroundings: while the photosphere's average temperature is an incandescent 5,800 degrees Celsius, sunspots may be a mere 3,000 degrees Celsius.

Sunspots form in pairs, marking regions where loops of magnetic field push out through the photosphere. Gas flowing along the loops can appear as dark filaments when silhouetted against the brighter background of the disc, or as pinkish prominences along the Sun's edge during an eclipse. The most spectacular solar activity of all occurs when magnetic loops 'short-circuit' and reconnect closer to the Sun. This process releases vast amounts of energy as radiation in solar flares that can boost the temperature of surrounding gases by millions of degrees, ejecting clouds of energetic particles into space at speeds of up to 1,000 kilometres per second (2,237,136 miles per hour).

Recent solar cycles
Cycle 25 (Dec 2019–)
Cycle 24 (Jan 2008–Dec 2019)
Cycle 23 (May 1996–Jan 2008)
Cycle 22 (Sept 1986–May 1996)
(Cycles are counted from 1755)

STARS **p.22** ORIGINS OF THE SOLAR SYSTEM **p.49** THE SUN IGNITES **p.50**
THE DEVELOPMENT OF THE SUN **p.55**

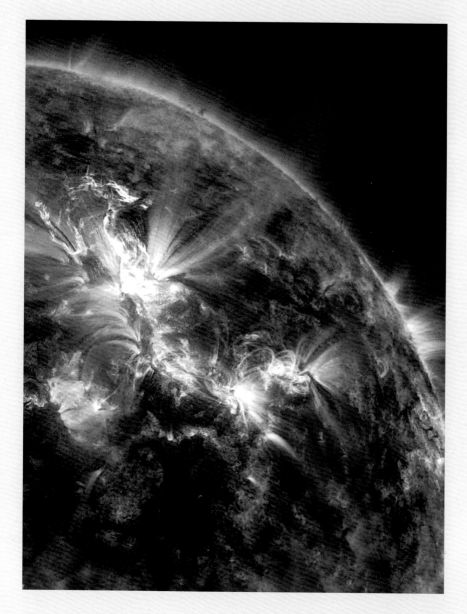

Flares above the Sun photographed by
NASA's Solar Dynamics Observatory.

Mercury

THE SOLAR SYSTEM'S SCORCHING INNERMOST, SMALLEST AND FASTEST-MOVING PLANET

The smallest and innermost planet, Mercury is 40 per cent larger than Earth's Moon, and superficially similar in appearance – weak gravity and the Sun's heat have kept it from forming a substantial atmosphere and as a result it has been scarred by bombardment from countless meteorites over billions of years. However, Mercury's outsized metallic core (thought to be a result of an impact that blasted away much of the surrounding rocky mantle layer during its formation) has also had important effects. Over hundreds of millions of years, it seems to have first expanded, cracking the overlying crust into blocks, and then shrunk inwards. The resulting jostling among the oversized blocks above it has created long cliff faces where some regions are pushed hundreds of metres above their neighbours – taller than most cliffs on Earth and often stretching for hundreds of kilometres. Heat from the core also seems to have powered sporadic volcanic activity until about a billion years ago.

Mercury races around the Sun in just 88 days and turns on its axis in two-thirds of its year. The combined effect is that most parts of the surface experience a complete day-night cycle only once every two Mercury years, and as a result the planet has the most extreme temperature variations of any planet, ranging from 425 degrees Celsius at midday to –195 degrees Celsius at night.

GORDON PETTENGILL
Mercury's rotation rate was discovered by Gordon Pettengill (1926–2021). In 1965 he used the giant Arecibo radio observatory in Puerto Rico as a radar, sending beams of radio waves towards Mercury and analysing the signals that returned. He discovered that the planet was not, as previously believed, synchronously tidally locked with the Sun, rotating once for each orbit, but in fact rotating every 55 days.

NON-LUMINOUS OBJECTS **p.24** ORIGINS OF THE SOLAR SYSTEM **p.49** CREATION OF THE PLANETS **p.51**
SOAKING UP THE DEBRIS **p.54**

Key features of Mercury
Caloris Basin (impact crater)
Beethoven Basin (impact crater)
Discovery Rupes (cliff scarp)
Chao Meng-Fu (possible ice-filled crater)
Weird terrain

Data returned in 2015 by NASA's
MESSENGER probe enabled the
creation of this mineral map of Mercury,
the surface of which has scarcely
changed in the last 3 billion years.

COLLISIONAL ACCRETION (SOLAR SYSTEM FORMATION) **p.199** PLANETARY MIGRATION **p.200**

Venus

EARTH'S BRILLIANT, CLOUD-CLOAKED INNER NEIGHBOUR

MAGELLAN MISSION

Most current maps of the Venusian surface were made by the Magellan mission (1989–94). This NASA orbiter carried a complex 'synthetic aperture' radar system, previously used by Earth-orbiting satellites, which pierced the clouds with radio waves. By analysing the reflected signals, Magellan could measure not only the height of the landscape below, but also its slope, roughness and reflectivity.

Venus is Earth's near-twin in terms of size, and often the closest planet to our own. Cloaked in a highly reflective atmosphere, it appears as a brilliant sight in Earth's skies, but its apparent beauty hides a hellish reality. Venus's atmosphere is dominated by toxic carbon dioxide and has surface temperatures of around 460 degrees Celsius. Coupled with sulfuric acid rainclouds and crushing atmospheric pressures, this makes Venus the most hostile world in the Solar System.

Armoured landers that have braved the surface and orbiting probes that have pierced the clouds with radar reveal Venus as a world shaped by intermittent volcanic eruptions far more extensive than those typically found on Earth, with signs of ancient impact craters long ago wiped out (and fewer of today's smaller meteorites able to reach the surface through the thick atmosphere). The differences between Venus and Earth are thought to originate from Venus's position slightly closer to the Sun (which it orbits in 225 days). Warmer temperatures caused the planet's early oceans to evaporate, robbing the planet of the lubricating effects that played a key role in breaking Earth's crust into fragmented tectonic plates. The absence of water also limited the absorption of carbon from the early atmosphere into rocks in the crust, leaving an excess of carbon dioxide to create an intense greenhouse effect.

Key features of Venus
Maxwell Montes (mountain belt)
Maat Mons (shield volcano)
Ishtar Terra (highland continent)
Aphrodite Terra (highland continent)
Baltis Vallis (canyon system)

NON-LUMINOUS OBJECTS **p.24** ORIGINS OF THE SOLAR SYSTEM **p.49** CREATION OF THE PLANETS **p.51**
SOAKING UP THE DEBRIS **p.54**

This enhanced-colour photograph of the thick, dense atmosphere of Venus was taken by NASA's Mariner 10 during its 1974 fly-by of the planet.

COLLISIONAL ACCRETION (SOLAR SYSTEM FORMATION) **p.199** PLANETARY MIGRATION **p.200**

Earth

THE LARGEST ROCKY PLANET – AND THE ONLY KNOWN ABODE OF LIFE

Major tectonic plates
African Plate
Antarctic Plate
Eurasian Plate
Indo-Australian Plate
North American Plate
Pacific Plate
South American Plate

Our home planet is the third from the Sun and the largest of all the Solar System's rocky worlds. It lies in a region of moderate solar radiation (at an average of 149.6 million kilometres/92.9 million miles from the Sun) that puts it squarely in the 'Goldilocks zone', where liquid water can persist on the surface for long periods without freezing or evaporating. Coupled with substantial gravity to trap an atmosphere, and significant internal heat to drive geological activity, this makes Earth uniquely suited to the evolution of life.

Earth's circular orbit and tilted axis give rise to a cycle of seasons that helps maintain a temperate climate. Extensive oceans, meanwhile, drive a cycle in which water is transferred between liquid, solid (ice) and vapour forms, moving from the seas to the atmosphere to the land and back, and shaping the landscape in the process. A large central core of molten iron and nickel generates a strong magnetic field that envelops the Earth, protecting it from the ravages of the solar wind and helping to retain both atmosphere and oceans. Heat escaping from the core (combined with effects from the water cycle transforming dense volcanic rocks into lighter ones through processes of weathering and erosion) has also caused Earth's outer crust to separate into dozens of tectonic plates – fragmented jigsaw pieces that float on the rocks of the mantle layer below, moving around the surface at rates of centimetres per year. New crust forms where plates are separating (mostly below the oceans); mountains are built where plates are crumpled together; and volcano chains are powered by heat escaping where plates descend back into the mantle and melt.

This complex environment provides plenty of opportunities for the energy-harvesting chemical reactions that are used by all forms of life. Evidence suggests that the first simple organisms appeared more than 4.2 billion years ago, and over billions of years these early bacteria absorbed carbon dioxide from the atmosphere and replaced it with

ALFRED WEGENER
The idea that Earth's crust is broken into slow-moving plates was first proposed in 1912 by Alfred Wegener (1880–1930) as an explanation for unusual correspondences between the shapes of coastlines and fossils found on what are now widely separated continents. Wegener's theory was neglected until the 1950s, when explorers found signs of new crust forming on the deep ocean floor.

ORIGINS OF THE SOLAR SYSTEM **p.49** CREATION OF THE PLANETS **p.51** BIRTH OF THE MOON **p.52**
SOAKING UP THE DEBRIS **p.54**

Earthrise over the Moon, photographed
from Apollo 8, 1968.

COLLISIONAL ACCRETION (SOLAR SYSTEM FORMATION) **p.199** PLANETARY MIGRATION **p.200**
PANSPERMIA **p.210**

Viewed from the International Space Station 300 kilometres (200 miles) above Earth, sunrise casts long shadows over a cloudy Philippine Sea.

the oxygen that more complex animals thrive on today. Larger and more advanced forms of life began to emerge around 600 million years ago, and have since shaped every aspect of the planet.

The Moon

EARTH'S HUGE SATELLITE – A BARREN WORLD WITH A COMPLEX HISTORY

REGINALD ALDWORTH DALY
The theory that the Moon originated in a giant impact was first proposed in 1946 by Reginald Aldworth Daly (1871–1957) as an alternative to the prevailing idea that the Moon broke away from a rapidly rotating newborn Earth. Daly's theory went unnoticed until the post-Apollo era, when geologists realized it could help to explain similarities and differences between Earth rocks and Moon rocks.

Earth's Moon is the largest satellite compared to its parent among any of the major planets. It is thought to have formed 4.5 billion years ago after a massive interplanetary collision threw debris into orbit around the Earth. Since then, its orbit has slowly spiralled outwards, until today it lies at an average distance of 384,400 kilometres (239,000 miles) from Earth, which it orbits every 27.3 days. Tidal forces raised by the Moon tug at Earth's oceans and create daily tides, while Earth's influence on the Moon has slowed its rotation so that it turns once with each orbit and keeps one face permanently toward the Earth.

At a quarter of Earth's diameter and with just over 1 per cent of its mass, the Moon's gravity is far too weak to trap an atmosphere, and so it has been bombarded by countless impacts throughout its history. Rock samples brought back by the Apollo astronauts suggest that the size and frequency of incoming asteroids reached a peak around 3.9 billion years ago, when a series of major impacts created huge basins across the Moon. As the rate of impacts then diminished, many of these were filled by dark volcanic lava that welled up from the lunar interior over the next few hundred million years. This has left our satellite with its familiar mix of dark, relatively unscarred seas or 'maria', and bright, heavily cratered highlands.

Biggest lunar seas
Oceanus Procellarum (Ocean of Storms)
Mare Frigoris (Sea of Cold)
Mare Imbrium (Sea of Showers)
Mare Fecunditatis (Sea of Fertility)
Mare Tranquillitatis (Sea of Tranquility)

ORIGINS OF THE SOLAR SYSTEM **p.49** CREATION OF THE PLANETS **p.51** BIRTH OF THE MOON **p.52**
SOAKING UP THE DEBRIS **p.54**

The dark side of the Moon as viewed
from the International Space Station in
2015. The craggy topography is caused
by asteroid collisions and volcanic
eruptions over billions of years.

COLLISIONAL ACCRETION (SOLAR SYSTEM FORMATION) **p.199** PLANETARY MIGRATION **p.200**

Near-Earth Objects

SMALL BODIES THAT WANDER THE INNER SOLAR SYSTEM

Although most of the Solar System's smaller objects are confined in distinct zones such as the main asteroid belt and the Kuiper Belt, there are still plenty of strays that wander on elongated orbits through both the inner and outer Solar System. Inwards of the asteroid belt these are known as Near-Earth Objects (NEOs). Most are asteroids – originally residents of the main asteroid belt before being kicked out due to close encounters, collisions, or the gravitational influence of Jupiter. A few, however, have sparse, gassy atmospheres that reveal them as icy comets that have become trapped in the inner Solar System. NEOs range in size from the 40-kilometre (25-mile) long asteroid Ganymed down to chunks of rock just a few metres across.

NEO orbits are classified according to their relationship with Earth's. Amor NEOs have orbits entirely outside of Earth's, while Atiras stay closer to the Sun. Atens and Apollos, meanwhile, are two types of 'Earth-crossers': despite their name, this does not necessarily mean that they are collision risks; their orbits are usually inclined, meaning that they will pass over or under Earth's orbit. While astronomers can carefully monitor NEOs for potential threats in the coming decades, collisions are inevitable in the long term as numerous past impacts (such as the one that wiped out the dinosaurs 66 million years ago) can attest.

Notable Near-Earth Objects
Asteroid 433 Eros
Asteroid 1566 Icarus
Asteroid 4179 Toutatis
Asteroid 99942 Apophis
Comet 2P/Encke

CARL GUSTAV WITT
The first Near-Earth Object was discovered by German astronomer Carl Gustav Witt (1866–1946) working at the Berlin Urania Observatory in 1896. The Amor-group asteroid, now known as 433 Eros, was captured on film by chance when it strayed into a two-hour photographic exposure Witt was using to measure the orbit of another asteroid.

NON-LUMINOUS OBJECTS **p.24** ORIGINS OF THE SOLAR SYSTEM **p.49** CREATION OF THE PLANETS **p.51**
SOAKING UP THE DEBRIS **p.54**

This photo of the asteroid Eros was created using six images (NEAR Shoemaker, 2000).

 COLLISIONAL ACCRETION (SOLAR SYSTEM FORMATION) **p.199** PLANETARY MIGRATION **p.200**
PANSPERMIA **p.210**

Meteorites

FRAGMENTS OF WORLDS BOTH LARGE AND SMALL, FOUND ON EARTH'S SURFACE

Largest known meteorites
Hoba (60 tonnes)
Cape York (31 tonnes)
Campo del Cielo (31 and 29 tonnes)
Armanty (28 tonnes)
Bacubirito (22 tonnes)

NON-LUMINOUS OBJECTS **p.24** ORIGINS OF THE SOLAR SYSTEM **p.49** CREATION OF THE PLANETS **p.51**
SOAKING UP THE DEBRIS **p.54**

As Earth makes its annual orbit around the Sun, it is constantly encountering smaller objects on their own distinct orbits. Most of this debris, such as dust particles from the tails of comets, burns up high in the upper atmosphere as meteors or shooting stars, but every year several thousand larger chunks of rock make it through the atmosphere to Earth's surface. Most of these meteorites will splash into the sea or fall in areas where they go unnoticed, but many are recovered by meteorite hunters with an eye to an out-of-place rock.

Meteorites can originate from a variety of sources. A few rare ones were ejected into space during impacts on the Moon or Mars, while others are the stony or metallic remnants of large asteroids that broke up long ago. Most, however, are samples of rock that has altered little since the beginnings of the Solar System. These 'chondrites' contain tiny mineral spheres that are direct fossils of material that coalesced in the solar nebula. While most of them have been altered by heat as they fused together into larger space rocks, a small subset known as carbonaceous chondrites are truly unaltered, and give scientists their most valuable insight into the balance of chemicals that produced our Solar System.

ERNST CHLADNI

Though best known for his study of acoustics, German physicist Ernst Chladni (1756–1827) was also the first person to develop a scientific theory linking fireballs in the sky to reports of iron-rich rocks falling to the ground (previously dismissed as folktales). Although initially ridiculed, Chladni's work inspired a more thorough investigation of fireball sightings that ultimately confirmed his ideas.

The 60-tonne Hoba meteorite hit Namibia around 80,000 years ago.

COLLISIONAL ACCRETION (SOLAR SYSTEM FORMATION) **p.199** PLANETARY MIGRATION **p.200** PANSPERMIA **p.210**

Mars

THE OUTERMOST ROCKY PLANET, AND OUR MOST EARTHLIKE NEIGHBOUR

Key features of Mars
Olympus Mons (shield volcano)
Valles Marineris (canyon system)
Tharsis Montes (volcano chain)
Hellas Basin (impact crater)
Vastitas Borealis (lowland plain)

The Red Planet is the outermost of the rocky major planets, and just over half the size of Earth. Its orbit is distinctly elongated, ranging between 1.38 and 1.66 AU from the Sun, making for a chilly climate with an average temperature around –60 degrees Celsius. Yet in many ways Mars is the most Earthlike world in the Solar System, with a day just half an hour longer and a similarly tilted axis giving rise to a familiar pattern of seasons. However, the thin carbon-dioxide atmosphere exerts just 1 per cent of Earth's atmospheric pressure. The terrain is covered with fine dust which is tinted red by the same chemical (iron oxide) present in rust. In some regions, short-lived, localized dust devils can sweep the dust into the air, scouring it away to reveal darker rocks beneath. When Mars is at its closest to the Sun, seasonal winds can create huge dust storms that block the planet's surface from view for many weeks before gradually subsiding.

The varied Martian geography shows signs of an active history until quite recently. The largest Martian volcanoes (including the Solar System's highest mountain, Olympus Mons, roughly three times higher than Mount Everest) began to form billions of years ago, but there are signs of lava floes on their flanks (and smaller volcanoes elsewhere on the planet) that seem to have been formed in the past few million years. Another prominent feature is the deep fault system known as the Valles Marineris, four times deeper than the Grand Canyon and some 4,000 kilometres (2,500 miles) long.

Evidence from orbiters and surface landers shows that Mars was once much warmer and wetter than it is today.

CURIOSITY MISSION
The most conclusive evidence for a warmer, wetter Martian past has come from NASA's Curiosity rover. Since 2012, this automobile-sized robot has explored Gale, an ancient impact crater that was once filled by a deep lake that left sediment layers on its floor. By analysing minerals in these rocks, Curiosity has opened a unique window onto the deep Martian past.

NON-LUMINOUS OBJECTS **p.24** ORIGINS OF THE SOLAR SYSTEM **p.49** CREATION OF THE PLANETS **p.51**
SOAKING UP THE DEBRIS **p.54**

A photomosaic of Mars composed from images sent back to
Earth by Viking Orbiter 1 in 1980. The channels on the surface
suggest that the planet once had abundant water.

COLLISIONAL ACCRETION (SOLAR SYSTEM FORMATION) **p.199** PLANETARY MIGRATION **p.200**
PANSPERMIA **p.210**

In 2018, cameras aboard the European Space Agency's Mars Express (launched in 2003) revealed seasonal winds whipping up short-lived Martian dust devils that scrawl crazy patterns as they clear away the light surface soil to reveal darker terrain beneath.

Dried-up river beds and signs of past catastrophic flooding are plentiful. While much of the ancient ocean and atmosphere has been lost to space (thanks in part to the planet's lack of a protective magnetic field), vast reserves are still locked away beneath upper layers of carbon dioxide

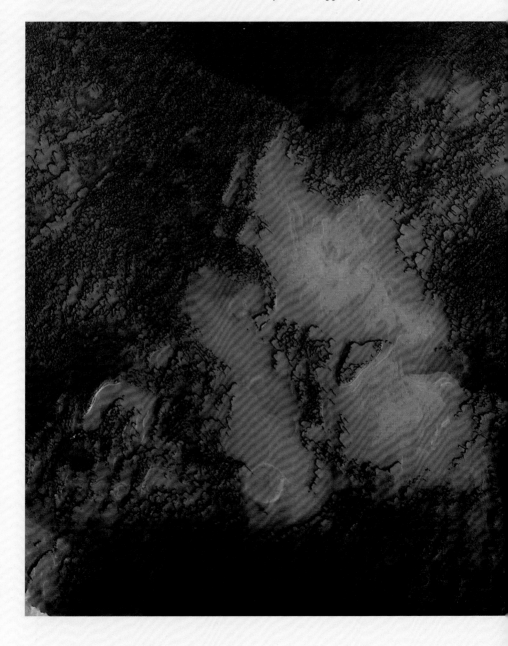

ice in the planet's polar ice caps, and mixed with the soil to form an extensive permafrost that creates glacier-like features across the planet's southern hemisphere.

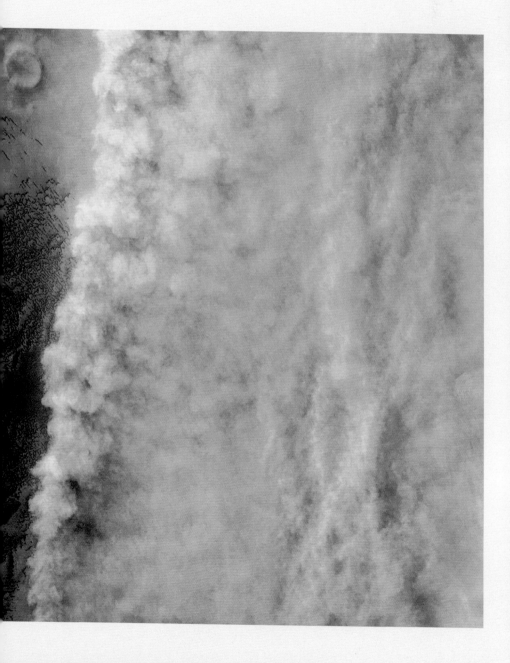

The Asteroid Belt

A CLOUD OF ROCKY WORLDS ENCIRCLING THE INNER SOLAR SYSTEM

The first asteroids
1 Ceres (1801)
2 Pallas (1802)
3 Juno (1804)
4 Vesta (1807)
5 Astraea (1845)
6 Hebe (1847)

Lying beyond the orbit of Mars but well inside that of the giant planet Jupiter, the Solar System's main asteroid belt is a doughnut-shaped region containing several hundred million mostly rocky objects. Around 200 have diameters greater than 100 kilometres (62 miles) – the very largest of all, Ceres, qualifies as a dwarf planet – but the combined mass of all this material is less than 5 per cent of the mass of Earth's Moon. The belt is thought to mark a region where Jupiter's gravitational influence prevented the formation of a fifth rocky planet, and the surviving asteroids are just a small fraction of the material that would originally have orbited in this region before being thrown towards the Sun or flung into the distant depths of the Solar System.

The asteroid belt's vast volume means that it is mostly empty space, with none of the jostling between asteroids depicted in science fiction films. Nevertheless, on million-year timescales, collisions and close encounters are regular occurrences, giving rise to distinct asteroid families with shared compositions and similar orbital features. Occasionally such encounters can still eject an asteroid from the belt – particularly if it strays into forbidden zones known as the Kirkwood gaps, where its orbital period will bring it under the influence of Jupiter.

JOHANN ELERT BODE
The existence of the asteroid belt was predicted by Johann Elert Bode (1747–1826), when he noted an apparent mathematical sequence in the size of planetary orbits. Bode's Law (which may be linked to the way solar systems form) predicted there 'should' be a planet between the orbits of Mars and Jupiter, and inspired astronomers to begin the search for it.

Artist's impression of the asteroid belt
between Jupiter and Mars. Nearly a
million of these rocky bodies are more
than 1 kilometre (0.6 miles) long.

COLLISIONAL ACCRETION (SOLAR SYSTEM FORMATION) **p.199** PLANETARY MIGRATION **p.200**

Ceres

THE LARGEST ASTEROID AND THE INNERMOST DWARF PLANET

The largest object in the asteroid belt, Ceres is classed by astronomers as a dwarf planet – a body that is massive enough to pull itself into a spherical shape under its own gravity, but does not exert enough influence to clear its surroundings of smaller nearby asteroids. It follows a somewhat elliptical orbit at an average of 2.8 AU from the Sun. This puts it beyond the Solar System's 'frost line' where ice can persist on the surface even in a vacuum – a significant boundary because objects beyond this distance tend to be formed from a mixture of rock and ice.

Images from the Dawn space probe have revealed that Ceres' surface is a dark mottled grey, heavily cratered with a few bright white spots. The crust appears to be a mix of ice and clay-like hydrated minerals, and the bright spots seem to be caused by briny salts seeping to the surface from a slushy mantle layer below, and then forming crystals as their water content evaporates. The evaporating water forms a tenuous atmosphere around Ceres that is constantly stripped away by particles from the solar wind – so the fact that the atmosphere persists today is key evidence that Ceres still has weak geological activity.

GIUSEPPE PIAZZI

While several of his contemporaries were deliberately searching for a 'missing' planet between Mars and Jupiter, Italian priest Giuseppe Piazzi (1746–1826) actually discovered Ceres, the first member of the asteroid belt, by chance. While compiling a detailed star catalogue he noticed a starlike object that moved slowly from night to night, but had none of the fuzziness associated with comets.

NON-LUMINOUS OBJECTS **p.24** ORIGINS OF THE SOLAR SYSTEM **p.49** CREATION OF THE PLANETS **p.51**
SOAKING UP THE DEBRIS **p.54**

Key features of Ceres
Ahuna Mons (ice volcano)
Kerwan (shallow impact basin)
Occator (impact crater)
Cerealia Facula (bright salt spot)
Oxo (crater with surface water)

From photographs taken in 2016 by
NASA's Dawn spacecraft, this image
shows varying levels of hydrogen
in Ceres' northern hemisphere: the
colour scale goes from blue (lowest)
to darkening reddish brown (highest).

COLLISIONAL ACCRETION (SOLAR SYSTEM FORMATION) **p.199** PLANETARY MIGRATION **p.200**

Vesta

A BATTERED ASTEROID WITH AN ACTIVE HISTORY

The third largest object in the asteroid belt, Vesta is very different from most of its siblings. Orbiting at an average of 2.4 AU from the Sun, it has a rockier composition than more distant, ice-enriched worlds, and a bright surface of volcanic rock, indicating that Vesta has seen far more geological activity than the majority of asteroids. With an average diameter of 525 kilometres (326 miles), it might have been large enough to pull itself into a spherical shape, were it not for the 500-kilometre (300-mile)-wide impact crater Rheasilvia that takes a vast chunk out of its south polar regions. Vesta is thought to be a rare surviving planetesimal – one of the small rocky bodies whose collisions built the planets of the inner Solar System. The rocky minerals that formed it contained enough radioactive material to heat its interior after formation, allowing heavy metals to sink inwards and form a nickel-iron core beneath a mantle of lighter, stony minerals. Heat escaping from the core as it cooled subsequently powered the volcanoes that re-drew much of the surface.

HEINRICH OLBERS
Vesta was first observed in 1807 by Heinrich Olbers (1758–1840). It was the fourth asteroid to be discovered. In 1802, Olbers had discovered Pallas, the second asteroid, and as evidence mounted that the new objects were faint and relatively small, he was the first person to argue for the existence of an entire asteroid belt.

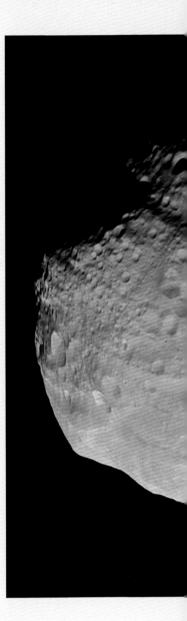

NON-LUMINOUS OBJECTS **p.24** ORIGINS OF THE SOLAR SYSTEM **p.49** CREATION OF THE PLANETS **p.51**
SOAKING UP THE DEBRIS **p.54**

Vesta as viewed by the Dawn spacecraft in 2011. The colour is unenhanced and natural, as it really appears.

Key features of Vesta
Rheasilvia Basin (impact crater)
Veneneia Basin (impact crater)
Feralia Planitia (eroded impact crater)
Divalia Fossa (canyon system)

COLLISIONAL ACCRETION (SOLAR SYSTEM FORMATION) **p.199** PLANETARY MIGRATION **p.200**

Jupiter

THE INNERMOST GAS GIANT AND THE SOLAR SYSTEM'S LARGEST PLANET

Major Jovian cloud features
Equatorial Zone
North Equatorial Belt
South Equatorial Belt
Great Red Spot
Oval BA (since 2000)

With a diameter more than 11 times Earth's and 318 times our planet's mass, Jupiter is the Solar System's biggest planet and the innermost gas giant – a world made mostly of lightweight hydrogen, but with a complex chemistry and climate that combine to produce colourful clouds.

Despite its vast size, Jupiter rotates on its axis in less than 10 hours, causing its equator to bulge outwards. The rapid spin wraps the planet's major weather systems parallel to its equator, creating alternating bands of light and dark clouds. Vast amounts of heat escape from within the planet as its inner regions contract due to gravity, driving high winds and violent storms including the famous Great Red Spot – an Earth-sized anticyclone that has been raging for centuries.

Orbiting the Sun every 12 years, Jupiter's enormous gravity exerts a powerful influence over nearby space, while its vast magnetic field stretches all the way to the orbit of Saturn. A sparse, dusty ring system and a vast family of moons are held in orbit by its gravity, while anything else that strays too close or even makes frequent alignments with Jupiter is likely to be disrupted. In this way, Jupiter has controlled the structure of the asteroid belt and even shaped the evolution of the Solar System.

JUNO MISSION
New insights into Jupiter's internal structure and complex weather have come from NASA's Juno mission, which entered orbit around the giant planet in 2016. The probe follows a unique orbit that is highly elongated and tilted so that it passes close to the planet over the poles but otherwise stays safely beyond its dangerous radiation belts.

ORIGINS OF THE SOLAR SYSTEM **p.49** CREATION OF THE PLANETS **p.51** ORDER OF THE PLANETS **p.53**
SOAKING UP THE DEBRIS **p.54**

Jupiter's southern hemisphere as imaged in 2020 by NASA's
Juno space probe. The swirling clouds are frozen hydrogen and
helium blown around by fierce solar winds.

COLLISIONAL ACCRETION (SOLAR SYSTEM FORMATION) **p.199** PLANETARY MIGRATION **p.200**

Io

JUPITER'S RADIATION-BLASTED, VOLCANO-WRACKED INNER MOON

LINDA MORABITO
Io's volcanic activity was discovered by NASA engineer and astronomer Linda Morabito (b.1953) during the Voyager 1 fly-by of 1979. While analysing a photograph taken of the crescent Io for navigation purposes, Morabito spotted a large but faint crescent hanging above the moon. This was eventually confirmed to be a huge plume of sulfur ejected by a volcano on Io's surface.

Jupiter's four largest satellites are known as the Galilean moons – complex worlds that might be considered planets in their own right were it not for their orbits. Io is the innermost of the four; a little larger than Earth's Moon and orbiting Jupiter at a similar distance (though Jupiter's powerful gravity means it completes each orbit in just 42.5 hours). Not only does Io's orbit take it through the doughnut-shaped belts of deadly radiation that are trapped by the giant planet's magnetic field, it also takes a pummelling from Jupiter's immense gravity. Tidal forces pull and knead its interior, producing huge amounts of heat as its rocks are stretched and pushed past each other. The heating effect has dried out Io's original icy component, and now powers abundant volcanic activity on the moon's surface, involving both easily melted, sulfur-rich compounds and Earth-like silicate lavas that require much higher temperatures to liquefy.

Io's volcanoes take various forms, ranging from vast plumes of sulfur erupting high into the sky (which contribute material to a distinct glowing ring of energized particles around Io's orbit), to towering cones and oozing lava craters. Eruptions lay down a multitude of different sulfur compounds, creating colourful deposits of red, orange, yellow, white and green that re-cover the moon's entire surface every few decades.

Key features of Io
Pele (plume-producing volcano)
Loki Patera (shield volcano)
Prometheus (lava flow)
Boösaule Montes (mountain chain)

ORIGINS OF THE SOLAR SYSTEM **p.49** CREATION OF THE PLANETS **p.51** SOAKING UP THE DEBRIS **p.54**

Io is the most volcanically active body in the Solar System. This enhanced-colour image of the moon comes from the Galileo Orbiter (1997).

COLLISIONAL ACCRETION (SOLAR SYSTEM FORMATION) **p.199** UNIVERSAL GRAVITATION **p.201**

Europa

AN ICE-COVERED MOON OF JUPITER WITH A DEEP HIDDEN OCEAN

Jupiter's second major moon is its smallest, but also its most intriguing. Like most moons of the outer Solar System, it is made from a combination of rock and ice, but in Europa's case (thanks to a weaker form of the same tidal heating that affects Io), internal warmth has caused these to separate significantly, creating a rocky core, an icy crust – and in between them a deep ocean of liquid water.

Outward evidence for Europa's hidden ocean comes from its brilliant white, highly reflective surface; remarkably smooth and almost entirely crater-free, it must have formed in the relatively recent past. Criss-cross tracks of 'dirty' ice that have apparently welled up from

Key features of Europa
Conamara Chaos (jumbled
terrain)
Rhadamanthys Linea (fracture
system)
Agenor Linea (fracture system)
Boeotia Macula (dark spot)
Pwyll (impact crater)

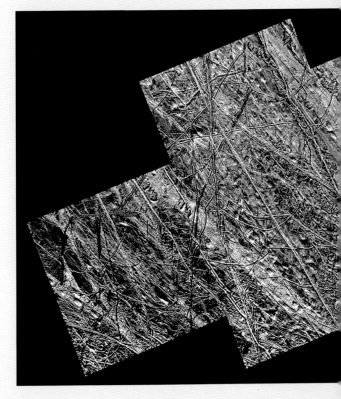

This mosaic of Europa's northern
hemisphere was created from images
sent by the US space probe Galileo
Orbiter between 1996 and 1998.

ORIGINS OF THE SOLAR SYSTEM **p.49** CREATION OF THE PLANETS **p.51** SOAKING UP THE DEBRIS **p.54**

the interior, coupled with signs of movement and collision between icy plates (the Europan equivalent of Earth's plate tectonics) offered an important clue, and later measurements of Europa's magnetic field confirmed the presence of a water layer up to 100 kilometres (62 miles) deep, from which warm ice and even liquid water well up to the surface. Experts think that the ocean is supplied with heat and enriched with chemicals by seafloor volcanoes – the similarity between this environment and Earth's own deep-sea vents, where many biologists think life first took hold, makes Europa the Solar System's most likely location for extraterrestrial life.

STANTON J. PEALE
The tidal heating effect that heats Io and Europa was predicted by Stanton J. Peale (1937–2015) mere weeks before the Voyager 1 fly-by of Jupiter in 1979. Building on earlier work, he applied a generalized law of tides to show how the shape of these moons would be distorted and flexed, generating friction and heat in their rocks.

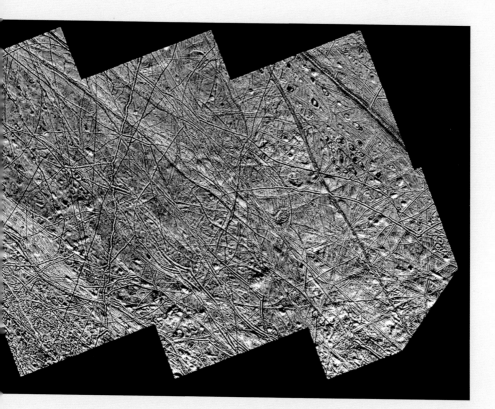

COLLISIONAL ACCRETION (SOLAR SYSTEM FORMATION) **p.199** UNIVERSAL GRAVITATION **p.201**
PANSPERMIA **p.210**

Ganymede

A GIANT SATELLITE OF JUPITER WITH A TORTURED PAST

The largest moon in the Solar System, at 5,269 kilometres (3,274 miles) across, Ganymede is larger than the planet Mercury. Made from a similar mixture of rock and ice to that of its immediate neighbours, Europa and Callisto, it has no atmosphere or signs of current activity. However, a patchwork surface of dark, heavily cratered areas and brighter, smoother ones, often separated by areas of parallel grooves named *sulci*, suggest that Ganymede has had an active past.

The properties of Ganymede's magnetic field suggest that its interior contains a core which is still warm enough to contain molten iron, and a briny ocean about 200 kilometres (124 miles) beneath the surface. While Ganymede does not currently suffer from the same tidal heating as the inner moons Io and Europa, it may have done so in the past when its orbit was slightly different, and this (coupled with radioactive heating from Ganymede's rocky minerals) seems to have allowed materials of different densities to melt and separate into distinct layers. Subsequently, heat from the core drove the giant moon's own version of plate tectonics, causing the ancient, cratered crust to fragment and rearrange itself, flooding some areas completely with fresh ice and creating the *sulci* in other areas where ice welled to fill in the fracture zones.

GALILEO

Jupiter's four large moons are generally known as the Galilean satellites, after Galileo Galilei (1564–1642). Galileo made many important discoveries and inventions, but is best known for his observations with an early self-built telescope. The discovery of moons circling Jupiter showed that not all celestial motion was centred on Earth, and inspired Galileo's support for the notion of a Sun-centred Universe.

Key features of Ganymede
Galileo Regio (dark plain)
Enki Catena (crater chain)
Tros (impact crater)
Memphis Facula (ice-filled crater)
Uruk Sulcus (grooved terrain)

Ganymede photographed by Juno as
the NASA spacecraft passed the moon
in 2021. The moon could have a briny
ocean under its surface.

COLLISIONAL ACCRETION (SOLAR SYSTEM FORMATION) **p.199**

Callisto

JUPITER'S HEAVILY BATTERED OUTER GIANT SATELLITE

Key features of Callisto
Asgard (impact basin)
Valhalla (impact basin)
Gomul Catena (crater chain)
Doh (crater with domed centre)

The outermost of Jupiter's major moons, Callisto is only slightly smaller than Ganymede, but the two worlds otherwise present a stark contrast. Callisto's dark surface is saturated with vast numbers of craters of all sizes, including the enormous ringed Valhalla impact basin, some 1,900 kilometres (1,200 miles) wide. Many craters have bright centres and surrounding 'rays', while at the heart of the largest basins are light and relatively smooth areas named palimpsests.

Callisto orbits Jupiter every 16.7 days, immune to the tidal forces that have heated and shaped its inner neighbours. As a result, its surface has been largely unaltered throughout its 4.5 billion-year history, although it has been hit by countless objects drawn to their doom by Jupiter's gravity. This has left Callisto as the most cratered body in the Solar System. Chemical changes triggered by particles and radiation from the Sun have caused the exposed surface of rock and ice to darken, but bright splashes form when fresh impacts expose and excavate ice from just beneath the surface. The bright, smooth palimpsest areas seem to have formed when fresh ice – perhaps linked to the briny ocean layer that still persists 150 kilometres (93 miles) beneath the surface – welled up long ago to heal the scars of major impacts.

GALILEO MISSION
Much of our understanding of Jupiter's moons comes from the Galileo space probe, which orbited Jupiter from 1995 to 2003. Galileo made numerous close encounters with the major moons, and its measurements of their interaction with Jupiter's powerful magnetic field led to the discovery that Ganymede and Callisto must have electrically conducting briny ocean layers beneath their crusts.

The Asgard impact crater on Callisto measures
1,600 kilometres (990 miles) in diameter.

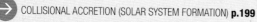

COLLISIONAL ACCRETION (SOLAR SYSTEM FORMATION) **p.199**

Saturn

A RINGED GIANT WITH A HIDDEN, STORMY ATMOSPHERE

Major Moons of Saturn
Mimas
Enceladus
Tethys
Dione
Rhea
Titan
Hyperion
Iapetus
Phoebe

Saturn is our Solar System's second gas giant – similar to Jupiter in terms of composition, but quite different in appearance. While it has less than one third of Jupiter's mass, its weaker gravity allows its hydrogen atmosphere to expand to almost the same size, giving Saturn an average density less than that of water (although the Earth-sized ball of rock and metal that probably forms its core is, of course, much denser). Accompanied by an abundant family of moons and circled by a spectacular system of icy rings, Saturn takes 29.5 Earth years to orbit the Sun at an average distance of 9.5 AU. Its polar axis is tilted just slightly more than Earth's, producing a similar cycle of seasons.

ORIGINS OF THE SOLAR SYSTEM **p.49** CREATION OF THE PLANETS **p.51** ORDER OF THE PLANETS **p.53**
SOAKING UP THE DEBRIS **p.54**

Saturn's weather systems, like Jupiter's, are wrapped into bands parallel to its equator but, in contrast to the colourful Jovian clouds, the colour palette of this planet appears mostly cream-and-white. In fact, the clouds where most of the action is happening are probably just as varied as Jupiter's, but their colours are muted by an overlying layer of white ammonia haze that forms in Saturn's cold upper atmosphere.

Although no permanent storms to rival the Great Red Spot of Jupiter have been found, some latitudes are capable of producing regular storm outbreaks at certain times in Saturn's long year.

PIONEER 11
Launched in 1973, NASA's Pioneer 11 was the first spacecraft to use a 'slingshot' manoeuvre to visit two planets. After using Jupiter's gravity to alter course during a 1974 fly-by, in 1979 it became the first probe to visit Saturn. The mission nearly met a premature end when it came within a few thousand kilometres of a previously undiscovered small moon.

This image of Saturn was taken in 2007 by the Cassini Orbiter. The planet is made mainly of hydrogen and helium.

COLLISIONAL ACCRETION (SOLAR SYSTEM FORMATION) **p.199** PLANETARY MIGRATION **p.200**

Saturn's Rings

THE SOLAR SYSTEM'S MOST SPECTACULAR RING SYSTEM

JAMES CLERK MAXWELL
The nature of Saturn's rings was first identified by physicist James Clerk Maxwell (1831–79). In an essay of 1859, Maxwell showed how variations in the strength of Saturn's gravity would cause a solid sheet of material to break apart, while a liquid would become unstable and form blobs. Therefore the rings could only be composed of particles in concentric orbits.

Saturn's ring system is one of the Solar System's most spectacular sights. Bright platters made up of countless particles following perfectly concentric, circular orbits stretch from the cloud tops to almost three times the planet's diameter (while more diffuse structures can be traced much further out).

The bright inner ring system comprises reflective chunks of ice in many different sizes. From the centre outwards, the tenuous D and semi-transparent C rings contain relatively small particles, while the opaque B and A rings consist of more substantial icy boulders. The rings contain several distinct divisions and gaps where ring particles are unable to remain in stable orbits due to the gravitational influence of Saturn's more distant large moons, while in other places, tiny 'shepherd moons' constrain the extent of thin rings like the F ring and act as sources for fresh material.

Despite their vast horizontal extent, the rings are remarkably thin, forming a plane just a few hundred metres deep. Their apparent solidity from a distance is entirely due to the vast numbers of particles they contain. Their origin is still poorly understood, but they probably formed from the breakup of an earlier icy object (a lost moon or a passing comet). Collisions between particles within the rings are constantly regenerating their material and exposing fresh, bright ice on the surface.

BOUND SYSTEMS **p.18**   ORIGINS OF THE SOLAR SYSTEM **p.49**   CREATION OF THE PLANETS **p.51**
SOAKING UP THE DEBRIS **p.54**

Saturn's A and B rings. The dark area that separates them – known as the Cassini division – is a gap of 4,800 kilometres (3,000 miles).

Major gaps in the rings
Colombo Gap, Maxwell Gap (within C Ring)
Huygens Gap (within Cassini Division)
Encke Gap, Keeler Gap (within A Ring)
Roche Division (between A and F Rings)

COLLISIONAL ACCRETION (SOLAR SYSTEM FORMATION) **p.199** UNIVERSAL GRAVITATION **p.201**

Titan

THE COLD AND COMPLEX GIANT MOON OF SATURN

Saturn's largest moon, Titan is the Solar System's most complex satellite. Larger than the planet Mercury, it retains a substantial atmosphere that is dominated by nitrogen, but rendered opaque by hazy orange methane clouds. Infrared cameras that can pierce these clouds have revealed a strangely Earth-like world below, with a mix of elevated 'landmasses' separated by low-lying, craterless plains. Many of the continental features appear smoothed and shaped by liquid erosion, despite surface temperatures of around –179 degrees Celsius.

Although no water can flow freely in Titan's frozen conditions, many of the satellite's features can be attributed to a 'methane cycle' – a low-temperature equivalent of Earth's water cycle in which the oily carbon compound methane shifts between vapour, liquid and solid states in the same way as water does on Earth.

Because Titan's atmospheric methane should be broken down by radiation from the Sun in just a few million years, its presence today suggests that it is being constantly replenished – probably by eruptions of methane-rich icy material through a handful of relatively warm volcanoes.

CHRISTIAAN HUYGENS

Titan was discovered in 1655 by Dutch physicist and inventor Christiaan Huygens (1629–95) using a self-built telescope. Huygens's instrument – a refracting (lens-based) telescope with a very long tube that could achieve a magnification of 50 times – was the most powerful of its time and also enabled him to identify the true shape of Saturn's rings.

BOUND SYSTEMS **p.18** ORIGINS OF THE SOLAR SYSTEM **p.49** CREATION OF THE PLANETS **p.51** SOAKING UP THE DEBRIS **p.54**

Key features of Titan
Xanadu (bright highland continent)
Shangri-la (dark lowland plain)
Kraken Mare (methane sea)
Sotra Facula (possible ice volcano)
Menrva (impact crater)

Infrared imaging aboard the Cassini Orbiter penetrated
the mists that envelop Titan to reveal the surface
of the moon which, after Ganymede, is the second
largest satellite in the Solar System.

COLLISIONAL ACCRETION (SOLAR SYSTEM FORMATION) **p.199** UNIVERSAL GRAVITATION **p.201**
PANSPERMIA **p.210**

Enceladus

AN INNER MOON OF SATURN WITH A HIDDEN OCEAN NEAR THE SURFACE

Although smaller than some of its neighbouring moons in the Saturn system, 504-kilometre (313-mile)-wide Enceladus is a stunning and complex world. Its brilliant white landscape and relatively small number of craters suggest a surface that is constantly being renewed.

Early photographs from the Voyager space probe fly-bys led astronomers to suspect that the surface of Enceladus was blanketed in fresh snow formed by water erupting through geysers from beneath the surface, and instantly freezing into fine ice crystals. This theory was confirmed in spectacular style when NASA's Cassini spacecraft flew directly through one of these geyser plumes as it escaped

THE CASSINI MISSION
NASA's largest and most ambitious single space probe, the bus-sized Cassini orbited Saturn from 2004 to 2017. During this time it sent back countless images of the planet, its rings and its moons. On two occasions it flew directly through the plumes erupting from Enceladus's south pole, detecting water, carbon dioxide and various carbon-based chemicals within them.

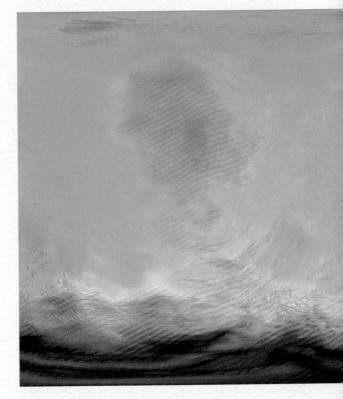

into space, thereby pumping new material into the tenuous
'E ring' that shares the moon's orbit around Saturn.

Although Enceladus's ice geysers were initially
assumed to contain a substantial amount of ammonia to
act as 'antifreeze' and make activity more viable at low
temperatures, Cassini's measurements showed that the
plumes are actually almost pure water. This means that
the moon probably has a global ocean layer, perhaps
warmed by tidal heating and other factors working
together. Such an ocean would make Enceladus an
intriguing candidate for the evolution of primitive life.
Enhanced-colour images reveal surface details including
bluish 'tiger stripes' near the south pole: these appear to
be the source of the current eruptions.

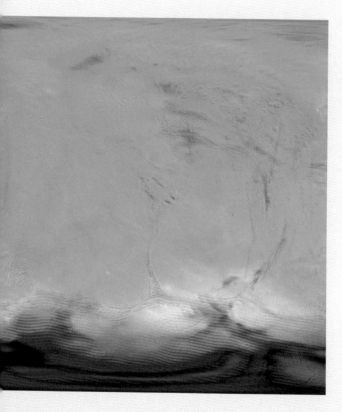

Key features of Enceladus
Labtayt Sulci (deep fracture
system)
Samarkand Sulci (grooved
'tiger stripe')
Sarandib Planitia (smooth plain)
Ali Baba (impact crater)
Dunyazad (impact crater)

Infrared map of Enceladus compiled
from data collected by the Cassini
Orbiter. The moon is covered in liquid
and frozen water; the newest deposits
of ice show up here in red.

COLLISIONAL ACCRETION (SOLAR SYSTEM FORMATION) **p.199** UNIVERSAL GRAVITATION **p.201**
PANSPERMIA **p.210**

Iapetus

SATURN'S CURIOUS MOON OF TWO HALVES

Key features of Iapetus
Cassini Regio (dark terrain)
Roncevaux Terra (light terrain)
Saragossa Terra (light terrain)
Turgis (impact basin)
Engelier (impact basin)

The third largest moon of Saturn, Iapetus is also the outermost of the satellites that formed alongside the giant planet instead of being captured into orbit later. This is a world of two halves, with a coal-black leading hemisphere that faces permanently in the direction of travel around its orbit, and a snow-white trailing half. The dark surface seems to lie on top of the brighter one, and there are few signs of the two terrains blending together. Nevertheless, both hemispheres are equally cratered.

Iapetus is thought to owe its remarkable appearance to the influence of the next satellite out – a large, dark object named Phoebe that is probably a large comet captured into orbit by Saturn's gravity. Dust from Phoebe spirals inwards until some of it is swept up by Iapetus's leading hemisphere. The dusty coating makes this side of the moon absorb more sunlight, warming its mix of rock and ice until the ice turns to vapour and escapes into space, leaving a dark rocky 'lag' behind, which itself helps to warm the hemisphere. In this way, a tiny initial difference between the hemispheres has become greatly exaggerated. Another intriguing feature of Iapetus is the huge equatorial ridge that wraps around much of the planet. Its origins are poorly understood, but at up to 20 kilometres (12 miles) high in places, it gives the moon a walnut-like appearance.

GIOVANNI DOMENICO CASSINI
Iapetus is one of four major moons of Saturn discovered by Giovanni Domenico Cassini (1625–1712) while director of the Paris Observatory. When he found that Iapetus was visible on one side of Saturn but not the other, he correctly deduced that the moon must have two contrasting hemispheres. The dark hemisphere is today called Cassini Regio.

BOUND SYSTEMS **p.18** ORIGINS OF THE SOLAR SYSTEM **p.49** CREATION OF THE PLANETS **p.51**
SOAKING UP THE DEBRIS **p.54**

The bright trailing hemisphere of
Iapetus. The impact crater that forms
its most prominent surface feature is
450 kilometres (280 miles) wide.

COLLISIONAL ACCRETION (SOLAR SYSTEM FORMATION) **p.199** UNIVERSAL GRAVITATION **p.201**

Uranus

AN ICE GIANT PLANET TILTED ON ITS SIDE

Rings of Uranus
Icy: 6, 5, 4, Alpha, Beta, Eta,
Gamma, Delta, Epsilon
Dusty: Zeta (inside ring 6),
Lambda (inside Epsilon)
Outer: Mu, Nu

Orbiting at roughly twice Saturn's distance from the Sun, Uranus is about half its size, and the first of the Solar System's two roughly identical 'ice giant' worlds. This distinctive class of planets is made up mostly of chemical 'ices' – not only water but also other chemicals with low melting points, such as ammonia and methane. Small amounts of methane in the atmosphere absorb some colours of sunlight and reflect others to give Uranus a pale blue-green colour.

During the only fly-by of Uranus in 1986, the planet appeared as an almost featureless ball, but since then Earth-based telescopes have occasionally captured more activity. Uranus's changing levels of activity may be linked to its very unusual climate and seasons: the planet's axis is tipped over at 98 degrees relative to its orbit, and as a result it 'rolls' around its 84-year orbit. Each hemisphere experiences a years-long winter of perpetual darkness, a period of relatively normal day and night, and then a summer of endless daylight. The planet's strange orientation is highlighted by its ring system, which surrounds Uranus like a target around a bullseye. The 13 rings are much narrower and more distinct than the broad planes around Saturn, and the ring material is darker – most likely due to a coating of reddish methane ice.

WILLIAM HERSCHEL
Uranus was discovered in 1781 by musician William Herschel (1738–1822) while surveying double stars with a home-made reflecting telescope. He first reported his discovery as a possible comet, but its slow movement through the sky soon revealed that the new object was much further away and much larger – a new planet so distant from its nearest neighbour, Saturn, that it doubled the extent of the Solar System.

ORIGINS OF THE SOLAR SYSTEM **p.49** CREATION OF THE PLANETS **p.51** ORDER OF THE PLANETS **p.53**
SOAKING UP THE DEBRIS **p.54**

The colour of Uranus is produced by
methane gas, which absorbs the red of
sunlight and makes the planet look blue.

COLLISIONAL ACCRETION (SOLAR SYSTEM FORMATION) **p.199** PLANETARY MIGRATION **p.200**

Uranian Moons

THE CURIOUS MOONS OF ICE GIANT URANUS

Uranus's 27 known moons follow a similar distribution to the much larger satellite families of Jupiter and Saturn. Close to the planet, some 13 small inner moons orbit in and around the ring system; beyond these lie larger 'major' satellites that formed from the same raw rotating clump of material as Saturn itself; finally the planet is accompanied by an outer cloud of small 'irregular' moons – captured comets and 'centaur' objects in tilted and elongated orbits.

Moving out from the planet, the major satellites are Miranda, Ariel, Umbriel, Titania and Oberon. Each contains roughly equal amounts of rock and ice, except for the smallest and innermost, Miranda, which is dominated by ice. All show signs of being shaped by an icy equivalent of Earth's plate tectonics at various times in their history – particularly broad canyons named chasmata, created as their original icy crusts cracked and shifted.

Miranda is the star of the show – its surface is a crazy jumble of mismatched terrains, including Verona Rupes, the tallest cliffs in the Solar System, 20 kilometres (12 miles) high. Upon first seeing it, planetary scientists assumed it must have been shattered into fragments and completely reassembled at some time, but today it is thought to be a victim of extreme tidal forces that caused its original surface to melt, collapse and re-form.

VOYAGER 2
So far, our only close-up glimpses of Uranus, Neptune and their moons are from a single spacecraft – NASA's Voyager 2. Along with its sibling Voyager 1, this robot probe launched from Earth in 1977. After flying past Jupiter in 1979 and Saturn in 1981, Voyager 2 then travelled on alone to encounter Uranus in 1986 and Neptune in 1989.

BOUND SYSTEMS **p.18** ORIGINS OF THE SOLAR SYSTEM **p.49** CREATION OF THE PLANETS **p.51**
SOAKING UP THE DEBRIS **p.54**

Key moon features
Verona Rupes, Miranda (cliffs)
Inverness Corona, Miranda (grooved terrain)
Ithaca Chasma, Ariel (canyon)
Wokolo, Umbriel (crater)
Messina Chasma, Titania (canyon)
Hamlet, Oberon (crater)

The fascinating and still mysterious
surface of Miranda has old, heavily
cratered terrain interspersed with lightly
cratered regions of parallel bright and
dark bands, scarps and ridges.

COLLISIONAL ACCRETION (SOLAR SYSTEM FORMATION) **p.199** UNIVERSAL GRAVITATION **p.201**

Neptune

THE SOLAR SYSTEM'S SURPRISINGLY ACTIVE OUTERMOST PLANET

Rings of Neptune
Galle Ring
Le Verrier Ring
Lassell Ring
Arago Ring
Adams Ring

The outermost major planet, Neptune orbits the Sun every 165 years, about 30 times further from the Sun than Earth. Like Uranus, it is an ice giant, and the two worlds are almost identical in size (a little less than four times the diameter of Earth). Aside from this, however, Neptune shows some important differences. Neptune's axis is tilted at an Earth-like angle far less crazy than that of Uranus, and so sunlight is far more evenly distributed in the course of its 16-hour day.

Neptune's deeper blue colour is due to methane and other unknown chemicals in its atmosphere, and the planet also displays far more violent and varied weather systems than Uranus. Its winds of up to 2,100 kilometres per hour (1,300 miles per hour) – the fastest in the Solar System – periodically generate large, dark storms in its atmosphere as well as bright high-altitude clouds.

Neptune's active weather is surprising considering how little heat it receives from the Sun – the temperature at its cloud tops is a feeble –218 degrees Celsius. However, the atmosphere is heated from within, because Neptune radiates about 2.6 times more energy than it receives. This is thought to be generated by a mix of gravitational contraction and chemical reactions taking place above the planet's rocky core.

URBAN LE VERRIER
As observers tracked Uranus's orbit after its 1781 discovery, they found it tended to drift from its expected location. Urbain Le Verrier (1811–77) attributed this to the pull of another undiscovered planet, and calculated its position in 1846. He sent his results to the Berlin Observatory, whose astronomers rapidly identified Neptune to within 1 degree of Le Verrier's predicted position.

ORIGINS OF THE SOLAR SYSTEM **p.49** CREATION OF THE PLANETS **p.51** ORDER OF THE PLANETS **p.53**
SOAKING UP THE DEBRIS **p.54**

Neptune viewed in 1998 from Voyager 2
at a distance of 7 million kilometres
(4.4 million miles) from the planet's
surface. Like Uranus, this planet is
composed mainly of ammonia, methane
and water.

COLLISIONAL ACCRETION (SOLAR SYSTEM FORMATION) **p.199** PLANETARY MIGRATION **p.200**

Triton

A LATE-ARRIVING MOON OF NEPTUNE WITH A DISRUPTIVE PAST

Key features of Triton
Leviathan Patera (volcanic dome)
Mahilani (geyser plume)
Dagon Cavus (cantaloupe depression)
Mazomba (impact crater)

This photograph of Triton, taken by Voyager 2 during its 1989 fly-by, shows ice formations and impact craters.

WILLIAM LASSELL
Triton was discovered a mere 17 days after Neptune itself, by merchant and amateur astronomer William Lassell (1799–1880). Lassell used a self-built instrument with an innovative mount that made it easier to track objects in the sky, and in addition to Triton also discovered Saturn's moon Hyperion and two moons of Uranus.

A moon of Neptune, Triton is the largest of that planet's 14 known satellites. Uniquely for any large moon in the Solar System, Triton orbits in the opposite direction to its planet's own rotation. For these reasons and more, Triton is thought to be an interloper – a rogue world from the Kuiper Belt region beyond Neptune, whose capture into orbit disrupted the original satellite system and scattered the original moons into deep space.

Triton's arrival at Neptune would have subjected the moon to tremendous tidal forces as its orbit was transformed to the perfect circle it is today. The heating effect of these tides led to melting of the interior and a widespread reconfiguration that covered much of the surface with rippled terrain that has been likened to the skin of a cantaloupe. Remarkably, despite a surface temperature of –235 degrees Celsius, Triton's internal heat is still driving geological activity today, in the form of nitrogen geysers that spew dust into a thin atmosphere and leave dark streaks across the surface.

BOUND SYSTEMS **p.18** ORIGINS OF THE SOLAR SYSTEM **p.49** CREATION OF THE PLANETS **p.51**
SOAKING UP THE DEBRIS **p.54**

Centaurs

SMALL ICY OBJECTS ORBITING AMONG THE GIANT PLANETS

Other notable centaurs
944 Hidalgo
7066 Nessus
8405 Asbolus
10199 Chariklo
32532 Thereus

Just as rocky Near-Earth Objects orbit between the inner planets, so the space between the giant planets is not empty: icy objects known as centaurs follow elliptical, often tilted orbits between just sunward of Saturn and just beyond Neptune. The first such object to be discovered, 218-kilometre (135-mile) Chiron, developed a gassy halo as it came closer to the Sun on its 50-year orbit. It is now generally assumed to be a large comet, with a dark blue-grey surface and substantial amounts of water ice.

In contrast, Pholus, the second centaur to be discovered, has a brighter and redder surface than Chiron, probably due to significant amounts of carbon-rich 'organic' chemicals on its surface. More objects of both kinds have since been discovered, and the stark division between the two types remains a puzzle. In general, however, Centaurs are believed to be 'stray' objects from the Kuiper Belt beyond Neptune. In their current risky orbits, they are unlikely to survive for long before being ejected again during a close encounter with a giant planet.

CHARLES T. KOWAL

Chiron, the first centaur, was discovered in 1977 by Charles T. Kowal (1940–2011) at California's Palomar Observatory. It was the most distant object found during his decade-long survey of the region around the plane of the Solar System. Its icy nature was not discovered until 1988, when it suddenly brightened by 75 per cent and developed a comet-like coma.

Artist's impression of Chiron with Saturn in the background.

COLLISIONAL ACCRETION (SOLAR SYSTEM FORMATION) **p.199** PLANETARY MIGRATION **p.200**

Pluto and Charon

THE LARGEST DWARF PLANET AND ITS GIANT MOON

The brightest and largest known member of our Solar System's Kuiper Belt, Pluto circles the Sun in a tilted and eccentric orbit that brings it closer to the Sun than Neptune for two decades of its 248-year duration, but takes it to almost 50 AU at its most remote. Pluto is about two-thirds the size of Earth's Moon, but despite its 2006 demotion to 'dwarf planet' status it has no fewer than five known satellites of its own. The largest of these, Charon, is more than half of Pluto's own diameter, and the two bodies are locked together by tidal forces so that one hemisphere of each world permanently faces towards the other.

Pluto's surface is a mix of starkly contrasting reddish-brown and brilliant white areas. The darker regions are heavily cratered and probably get their colour from tarry chemicals that have developed through the long action of weak solar radiation on frozen methane. The bright areas appear to have been created by much more recent resurfacing and are covered in nitrogen ice (the frozen form of the gas that makes up Pluto's meagre atmosphere). Glacier-like features suggest that this ice can flow slowly across the surface, while elsewhere there are huge mountains made of frozen water.

CLYDE TOMBAUGH

Pluto was discovered by Clyde Tombaugh (1906–97) during a deliberate but misguided search that struck lucky. Working on the then widespread assumption that Uranus and Neptune were being influenced by a further unseen planet, Tombaugh was employed to carry out a detailed photographic search at Arizona's Lowell Observatory. He found Pluto within months of beginning work.

Key features of Pluto

Tombaugh Regio (light plain)
Sputnik Planitia (ice-filled basin)
Hillary & Tenzing Montes (ice mountains)
Cthulu Macula (dark terrain)

ORIGINS OF THE SOLAR SYSTEM **p.49** CREATION OF THE PLANETS **p.51** ORDER OF THE PLANETS **p.53**
SOAKING UP THE DEBRIS **p.54**

Increasingly detailed photographs of Pluto, such as
this enhanced-colour image, are gradually revealing
the dwarf planet's complex and varied geological
history – and the Solar System's largest glacier.

COLLISIONAL ACCRETION (SOLAR SYSTEM FORMATION) **p.199** PLANETARY MIGRATION **p.200**

The Kuiper Belt and its Members

A BAND OF ICY WORLDS BEYOND THE MAJOR PLANETS

EDGEWORTH AND KUIPER
Despite being widely named after Dutch-American astronomer Gerard Kuiper (1905–73), the idea of a belt of icy worlds formed from the debris left beyond Neptune was first put forward by Irish amateur Kenneth Edgeworth (1880–1972) in 1943, eight years before Kuiper. It was not until 1992 that Albion, the second Kuiper Belt Object (after Pluto) was discovered.

Beginning around the orbit of Neptune, the Solar System is ringed with a cloud of icy worlds of which Pluto is the largest member so far discovered. This doughnut-shaped Kuiper Belt is mostly concentrated between 30 AU and 50 AU, beyond which it rapidly diminishes. The objects orbiting in the main belt range from substantial dwarf planets (not only Pluto, but also worlds named Haumea and Makemake, among others) down to small, icy comets. Most of them are thought to have formed close to their current orbits, though today's Kuiper Belt is probably just a shadow of its former self – according to some estimates 99 per cent of the bodies it once contained were destroyed or ejected from the Solar System as the giant planets shifted their orbits early in their history.

Little is known about individual Kuiper Belt Objects (KBOs) due to their great distance from Earth – the exceptions are Pluto and the small, dumb-bell-shaped Arrokoth, both of which have been visited in spacecraft fly-bys. Observations of KBOs using Earth-based telescopes have shown that they have a mix of dark grey and brighter red surfaces. One theory is that the colour comes from hydrogen sulphide ice – a chemical that could remain frozen on the colder, more remote KBOs, while evaporating into space from warmer ones slightly closer to the Sun.

Notable KBOs
16760 Albion (c.140 kilometres/87 miles)
20000 Varuna (c.660 kilometres/410 miles)
50000 Quaoar (1,121 kilometres/696 miles)
136108 Haumea (2,100 × 1,074 kilometres/
1,304 × 667 miles, dwarf planet)
136742 Makemake (c.1,480 kilometres/
920 miles, dwarf planet)

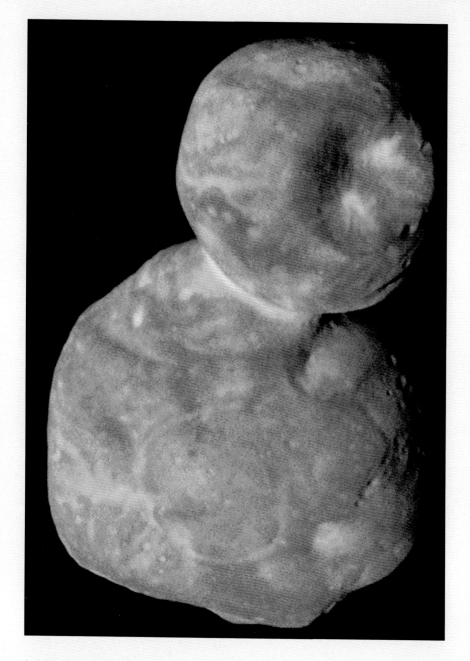

Arrokoth is 36 kilometres (22 miles) long
and 21 kilometres (13 miles) wide.

 COLLISIONAL ACCRETION (SOLAR SYSTEM FORMATION) **p.199** PLANETARY MIGRATION **p.200**

Eris and the Scattered Disc

DISTANT WORLDS EJECTED FROM THE KUIPER BELT

Slightly smaller in diameter than Pluto, the remote dwarf planet Eris is a member of the Kuiper Belt orbiting the Sun every 559 years. Its highly elliptical orbit takes it from 38 AU out to 98 AU from the Sun, and is sharply tilted at an angle of 44 degrees to the plane of the Solar System. This makes Eris the largest known object in the so-called 'scattered disc' – a sparse halo of objects on the fringes of the main doughnut-shaped Kuiper Belt. Objects in the scattered disc are generally thought to have started out in the main belt before being flung into their current orbits by close encounters with Neptune.

Despite its smaller size, Eris is 27 per cent heavier than Pluto – a fact revealed by the orbit of its small moon Dysnomia. This means that it must contain considerably more rock than other objects in this region of the Solar System. Some astronomers think the planet's rocky interior might also keep it warm enough to retain a liquid ocean layer beneath its surface, and perhaps even some form of geological activity. Another unusual feature is Eris's bright grey surface (a striking contrast to the dark reds and browns of its neighbours), which is thought to be a result of fresh methane frosts.

MIKE BROWN

When Eris was discovered in 2005 by Mike Brown (b.1965) and his colleagues, it was hailed as the Solar System's tenth planet. The likely discovery of further Eris-sized worlds, however, led astronomers to introduce the concept of 'dwarf planets', including both Pluto and Eris. Brown continues to search for a new major planet in the region beyond the scattered disc.

Notable scattered disc objects
$1996TL_{66}$ (575 kilometres/357 miles)
136199 Eris (dwarf planet, 2,326 kilometres/1,445 miles)
$2004XR_{190}$ (560 kilometres/348 miles)
225088 Gonggong (likely dwarf planet, c.1,230 kilometres/764 miles)

Artist's impression of the surface of Eris,
which is now known to be composed
mainly of frozen methane.

COLLISIONAL ACCRETION (SOLAR SYSTEM FORMATION) **p.199** PLANETARY MIGRATION **p.200**

Comets

DIRTY SNOWBALLS THAT CAN PUT ON A BRILLIANT DISPLAY

Bright historic comets
Great September Comet (1882)
Daylight comet of 1910
Comet Ikeya-Seki (1965)
Comet Hale-Bopp (1996)
Comet McNaught (2007)

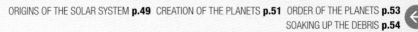

Comets can be spectacular but fleeting visitors to the inner Solar System. They are mountain-sized chunks of ice and rock whose surface begins to vaporize as they approach the heat of the Sun, forming a planet-sized atmosphere called a coma and potentially a tail millions of kilometres long as their gases are caught up on the solar wind and blown away from the Sun.

However, these beautiful visitors account for only a tiny minority of all comets – most of them remain in dormant deep-freeze within the Kuiper Belt and the Oort Cloud at the outer reaches of the Solar System (from where the active ones return at the outer limits of their long elliptical orbits). It takes a chance collision or the gravitational influence of a larger body to send a comet drifting sunwards, and each passage past the Sun drives off more of the ice that keeps it active. Often a 'long-period' comet from the outer Solar System may have its orbit dramatically altered by an encounter with one of the giant planets, either flinging it out of the Solar System entirely or shortening the period between returns to a few decades or even a few years. Many of these comets are ultimately destroyed in collisions with Jupiter, while those that survive will eventually exhaust their ice completely and become outwardly indistinguishable from rocky asteroids.

EDMOND HALLEY
The idea that comets could make predictable returns was first put forward in 1705 by Edmond Halley (1656–1742). Using his friend Isaac Newton's laws of motion and gravitation, he predicted that comets seen in 1456, 1531, 1607 and 1682 were one and the same, and furthermore that the comet (now named after him) would return in 1758.

Comet McNaught photographed over the Australian Outback, 2007.

COLLISIONAL ACCRETION (SOLAR SYSTEM FORMATION) **p.199** PLANETARY MIGRATION **p.200**
PANSPERMIA **p.210**

Sedna and the Oort Cloud

**ICY WORLDS AND DORMANT COMETS ORBITING AT THE EDGE
OF THE SUN'S GRAVITATIONAL INFLUENCE**

Artist's impression of Sedna looking
towards the Sun.

JAN OORT

Convincing arguments for a distant comet cloud were first put forward in 1950 by Jan Oort (1900–92). He showed that, as well as acting as an origin for long-period comets, such a cloud was needed to replenish the shorter-period comets. Without renewal, these objects would inevitably fade over a few million years, or have their orbits disrupted by planetary encounters.

Far beyond the Kuiper Belt, at the limits of the Sun's gravitational grasp, the Solar System is surrounded by the Oort Cloud – a huge hollow sphere more than a light year across of trillions of deep-frozen comets. The cloud cannot be seen, but astronomers learned of its existence by tracing back the orbits of the longest-period comets to their outermost points – rare visitors sent falling towards the Sun by occasional collisions and even the gravitational tides raised by other stars passing nearby.

The comets orbiting here began their lives much closer to the Sun before being thrown into their current orbits by encounters with Jupiter's powerful gravity early in their history. Around the plane of the Solar System, a disc-shaped inner Oort Cloud extends closer to the Sun. The most distant Solar System objects currently known, such as Sedna, may spend the outer part of their orbits in this region. Discovered in 2003, Sedna is a Ceres-sized world with a brilliant red surface that takes about 11,400 years to orbit the Sun. It is about 76 times further from the Sun than Earth at its closest approach, but retreats to more than 12 times that distance at its most remote.

Furthest Solar System objects
$2015RX_{245}$ (255 kilometres/158 miles)
90377 Sedna (995 kilometres/618 miles)
$2013SY_{99}$ (202 kilometres/125 miles)
$2015KG_{163}$ (101 kilometres/63 miles)

→ COLLISIONAL ACCRETION (SOLAR SYSTEM FORMATION) **p.199** PLANETARY MIGRATION **p.200** PANSPERMIA **p.210**

Theories

The Big Bang Theory

KEY SCIENTISTS: STEPHEN HAWKING • JIM PEEBLES • MARTIN REES • ROGER PENROSE

KEY DEVELOPMENT

Part of the mystery of the Big Bang is the question of what occurred at the singularity, which our current mathematics cannot describe. However, in 2020 Roger Penrose (b.1931) won the Nobel Prize in Physics for his work with Stephen Hawking (1942–2018) on modelling gravitational singularities. If a theory of quantum gravity can be developed in the future, it may tell us about the conditions in the singularity that sparked the Universe.

The Universe was born 13.8 billion years ago in an event we call the Big Bang. It created space and time as we know them, starting out as a microscopic, infinitely dense point smaller than a proton that we call the Big Bang singularity. It contained all the matter and energy that we see in the Universe today.

Don't think of the Big Bang as an explosion in the conventional sense, with a point of origin exploding into a pre-existing volume, since nothing existed for it to erupt into. Instead, the way to think about the birth of our Universe is that it happened everywhere: that microscopic point expanded to become today's cosmos.

There are several pieces of evidence that back up the theory. The Cosmic Microwave Background radiation is the 'afterglow' of the Big Bang, the residual heat from that intense moment. Space is also expanding, implying that our cosmos must have been smaller in the past. If we could rewind time, we would see the Universe contract back down to the singularity.

Stephen Hawking, one of the pioneering scientists who worked on the Big Bang Theory.

Cosmic Inflation

KEY SCIENTISTS: ALAN GUTH • ANDREI LINDE • ALEXEI STAROBINSKY • PAUL STEINHARDT

Alan Guth, who devised the
original theory of inflation.

KEY DEVELOPMENT
It is thought that inflation
was driven by a quantum
field called the Inflaton Field,
and that when inflation
stopped, the field decayed
and dumped its energy into
making matter. Eternal Inflation
solves the problem of why
inflation stopped, but has the
consequence that parts of the
Universe must keep inflating
forever, with patches budding
off to form new universes all
the time, creating a multiverse.

Inflation is the theory that, when the
Universe was unimaginably young – just
10^{-35} seconds old – it underwent a brief but
rapid burst of expansion, swelling in size
from subatomic to macroscopic.

While the possibility that inflation really
occurred happens to fit the observational
facts – one side of the Universe looks the
same as the other, and the whole structure
seems remarkably smooth on large scales
– the theoretical understanding of why
inflation occurred, what drove it and why it
stopped is a little threadbare.

Indeed, the original theory of inflation
had trouble getting it to stop – it suggested
that inflation would not stop everywhere
at once, but in little pockets that would

then expand away from each other. So
in the 1980s this was adapted into a new
theory of inflation, called Eternal Inflation.
This suggests that inflation was patchy in
where it stopped and where it didn't, but on
much larger scales, and one of the patches
in which it did stop became our entire
visible Universe.

← THE MOMENT OF INFLATION **p.36**

Special Relativity

KEY SCIENTISTS: ALBERT EINSTEIN • HENDRIK LORENTZ • PAUL LANGEVIN

Imagine two cars racing. A stationary observer at the side of the racetrack might see one car speed past at 100 kilometres per hour (62 miles per hour) and the other at 90 kilometres per hour (56 miles per hour) relative to the observer's motion of 0 kilometres per hour. However, relative to the two cars, one car is travelling 10 kilometres per hour (6 miles per hour) faster – or slower – than the other.

Einstein imagined a similar scenario, but involving light beams. The speed of light is 299,792 kilometres (186,282 miles) per second. If a spaceship were travelling at half that speed while chasing a light beam, would its occupants then see the light beam moving at only half the speed of light?

Apparently not, said Einstein. He showed that the speed of light in a vacuum is the same for all observers, regardless of the observer's motion.

If you travel at the speed of light you will experience time normally, but observers moving more slowly will see time pass more slowly for you than for themselves. This effect is known as time dilation.

KEY DEVELOPMENT
Physicist Paul Langevin (1872–1946) showed that time dilation leads to something called the Twin Paradox. Imagine you have a twin, who goes off in a light-speed spaceship to travel around the Universe for 20 years according to clocks on Earth. The twin's clock would run slower relative to yours, so while you age a couple of decades, your sibling would barely age at all. When you meet again, your twin would be much younger than you.

Albert Einstein showed that light beams always travel at the same speed in a vacuum, and nothing can travel faster.

FUNDAMENTAL FORCES **p.30**

General Relativity

KEY SCIENTISTS: ALBERT EINSTEIN • ARTHUR EDDINGTON • KARL SCHWARZSCHILD ALEXANDER FRIEDMANN

Albert Einstein's famous theory from 1915 builds on his Special Theory of Relativity, but now introduces mass and gravity into the mix. His Theory of General Relativity explains how mass and energy are equivalent (as described in his equation, $E=mc^2$), and how they exert influence on space and time through gravity. General relativity unifies both space and time into a framework that we call spacetime, through which mass moves and, in turn, bends these four dimensions to its will. The more massive an object, the more it warps spacetime, and the more strongly curved spacetime is, the stronger the gravity in that area of the Universe is. We can see this in the case of black holes, and the weird effects that occur near the black hole's event horizon.

The Theory of General Relativity is currently our best law for explaining the Universe. However, it falls apart at the smallest scales where quantum physics dominates, and so in the future we will need a quantum theory of gravity to fully understand the birth of the cosmos.

KEY DEVELOPMENT

On 14 September 2015, the LIGO gravitational wave experiment detected ripples in spacetime coming from the merger of two black holes. Gravitational waves are predicted by general relativity as disturbances in spacetime created by the interaction of massive bodies. Their discovery – and the dozens of other gravitational-wave events found since – provided strong verification for the validity of Einstein's theory.

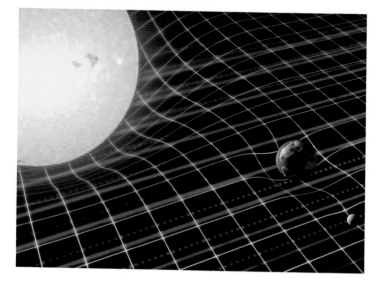

Mass is able to bend space and time; this is a key concept of general relativity.

 SPACETIME **p.14** STELLAR BLACK HOLES **p.122**

Multiverse

KEY SCIENTISTS: HUGH EVERETT • ERWIN SCHRÖDINGER • BRYCE DEWITT
ANDREI LINDE • MAX TEGMARK

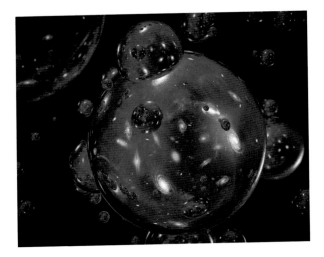

Computer illustration of multiple 'bubble' universes as predicted by the Eternal Inflation theory.

One possible and very weird consequence of quantum physics is that everything that can happen, does happen – but somewhere else, in another universe.

Quantum physics tells us that, at the smallest scales, the Universe is inherently fuzzy. A particle doesn't exist in one precise location. Instead, there is a range of possibilities describing the location where a particle could exist here. When we draw this probability distribution on a graph, it looks like a wavy line, which is called a wave function.

However, there is a school of thought that a particle can occupy every position on this wave function simultaneously in different parallel universes. This has led to the advent of the 'Many Worlds' theory, which predicts that a multiverse of universes exists.

There are other multiverse theories. In Eternal Inflation, inflation never ends in most of the Universe, but there are pockets where it does end and new universes are able to burst out, creating an infinite number of bubble universes. With this in mind, there could be an infinite number of versions of you reading this very book!

KEY DEVELOPMENT

The multiverse is a controversial idea, because it is untestable – we cannot access parallel universes to determine if they are real. For some scientists, this puts multiverse theories in the realm of pseudoscience, or even faith-based dogma. The existence of the multiverse would also mean that finding a theory of everything for our universe might be impossible, since anything we regard as a constant could have a different value in another universe.

THE MOMENT OF INFLATION **p.36** THE FATE OF THE UNIVERSE **p.59**

Collisional Accretion (Solar System Formation)

KEY SCIENTISTS: HAL LEVISON • EIICHIRO KOKUBO • SHIGERU IDA

The planets, including Earth, didn't just appear out of nowhere. They were built in what are known as protoplanetary discs – swirling pancakes of gas, dust and rocky rubble.

There are two sources of information that scientists can base their planet-forming models on. One is studying asteroids, comets and meteorites in our Solar System, which contain material untouched since the birth of the Solar System. The other is studying planet-forming discs around other stars – the more of them we can observe, the more complete a picture astronomers are able to build.

The key challenge at present is to understand how, once dust grains accrete into pebble-sized debris, they are then able to collide and merge to form larger objects without being smashed apart as they crash into each other. Some theories suggest that water-ice may have a role in sticking the pebbles together. Once these pebble aggregates gain enough mass, they can begin to soak up more and more material, building up step by step into protoplanets.

KEY DEVELOPMENT

The collisional accretion of the planets went in stages. After dust grains and pebbles have built protoplanets about a kilometre across, the protoplanets enter a stage of runaway accretion where their size jumps up to about 1,000 kilometres (1,600 miles) across in just 100,000 years. There may be several hundred large protoplanets, which then grow more slowly in a process called 'oligarchic accretion'.

The planets formed when smaller bodies – asteroids and protoplanets – collided and stuck together.

CREATION OF THE PLANETS **p.51** BIRTH OF THE MOON **p.52** SOAKING UP THE DEBRIS **p.54**

Planetary Migration

KEY SCIENTISTS: KEVIN WALSH • HAL LEVISON • ALESSANDRO MORBIDELLI
KLEOMENIS TSIGANIS

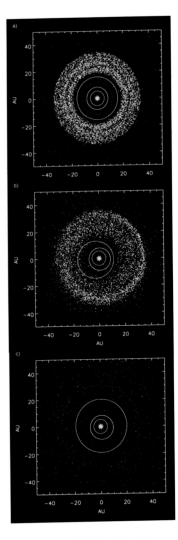

The giant planets originally formed in neatly circular orbits (top), but as they began to migrate they scattered smaller bodies – asteroids and comets – around the Solar System.

There are several odd things about the Solar System that could be neatly explained if the orbits of some of the planets wandered – or migrated – early in their history. The mass of Mars, for example, is less than might be expected, while the Asteroid Belt appears severely depleted. And a 'scattered disc' of icy objects beyond Pluto and the Kuiper Belt is full of icy comets on elongated objects that seem to have been thrown out there.

If the planets, particularly massive Jupiter and Saturn, were marauding around the early Solar System, then their gravitational influence could explain many of these oddities. We can see that migration occurred in exoplanetary systems. Migration happens when a giant planet grows larger by sweeping up gas as it moves around its star, thus creating a gap in the protoplanetary disc through which the planet is then able to migrate. The migration stops when the gas disc disappears, blown away by the stellar wind after a few million years.

KEY DEVELOPMENT

One of the leading candidates to explain planetary migration in the Solar System is the 'Grand Tack' model, developed by Kevin Walsh of the Southwest Research Institute in Colorado, USA. It describes how a young Jupiter, just a few million years old, migrated inwards, getting as close as 225 million kilometres (140 million miles) from the Sun; the gravity of Saturn brought its wandering to a halt before the duo returned to the outer Solar System.

JUPITER **p.154** SATURN **p.164**

Universal Gravitation

KEY SCIENTISTS: ISAAC NEWTON • HENRY CAVENDISH • ALBERT EINSTEIN

It was Isaac Newton (1642–1726) who first developed a scientific theory for gravity. Although the story of him being hit on the head by a falling apple may be untrue, it was the objects in the heavens that inspired him. And while his theory of gravitation has since been superseded by general relativity, for most scenarios in the Solar System his classical theory of gravity is sufficient to explain the motions of the planets, moons and comets.

Newton explained gravity mathematically as an attractive force that exists between all particles of matter, and he called this his Law of Universal Gravitation. The gravitational force between any two objects is dependent upon their mass and the distance between them – the closer they are, the stronger the force.

Newton's gravitational mathematics developed Johannes Kepler's (1571–1630) laws of planetary motion, which explained elliptical orbits.

KEY DEVELOPMENT
The first hint that there was more to gravity than Newton's law of universal gravitation was the orbit of Mercury, which Newtonian gravity could not explain. Astronomers invoked the existence of an unseen planet inside the orbit of Mercury, called Vulcan, which they argued was pulling on Mercury, but Vulcan was an illusion. Mercury's orbit, in the stronger gravitational field near the Sun, can be explained by general relativity.

Sir Isaac Newton developed the first theory of gravity.

FUNDAMENTAL FORCES **p.30** MERCURY **p.130**

Stellar Spectroscopy

KEY SCIENTISTS: NORMAN LOCKYER • JOSEPH VON FRAUNHOFER • EJNAR HERTZSPRUNG
HENRY NORRIS RUSSELL

If you pass white light through a prism, it will split into a swathe of different colours: red, orange, yellow, green, blue, indigo and violet. Each colour corresponds to a wavelength.

Stars also create this rainbow of light, known as a continuous spectrum, but they don't emit equally at all wavelengths. Hotter stars produce more shorter-wavelength light, such as blue and ultraviolet, while cooler stars emit more in longer wavelengths, such as red and infrared light. The Sun falls somewhere in the middle – and so we call it a yellow star.

Since a star's temperature is related to its mass and age, we can determine these properties by looking at its colour. This information can be plotted on the Hertzsprung–Russell (HR) diagram, a graph that charts the temperature, or colour, of stars against their luminosity, and tells us what stage of its evolution a star has reached – whether it is still burning hydrogen, or has evolved into a red giant, a supergiant, or a white dwarf.

KEY DEVELOPMENT
The second most common element in the Universe – helium – wasn't discovered until 1868, and it was found first on the Sun (it wasn't found on Earth until 1895) by the English astronomer Norman Lockyer (1836–1920), who discovered it spectroscopically from his observatory in Sidmouth, Devon, UK.

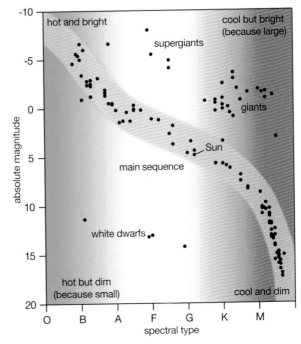

The Hertzsprung–Russell diagram of stellar evolution.

ELEMENTS **p.26** FORGING THE ELEMENTS **p.39** MAIN-SEQUENCE STARS **p.98**

Redshift and the Doppler Effect

KEY SCIENTISTS: VESTO SLIPHER • EDWIN HUBBLE • ADAM RIESS • GEORGES LEMAÎTRE

When a police car speeds past with its siren on, you hear the siren's pitch change. This is a manifestation of the Doppler Effect. As the vehicle comes towards you, sound waves are compressed, shortening their wavelength and raising their pitch. As it zooms away, the sound waves are stretched, increasing their wavelength, dropping their pitch.

Light can also act like a wave. The light of a galaxy moving towards us is compressed to shorter, bluer wavelengths. We call this blueshift. The light of galaxies moving away from us is stretched to longer, redder wavelengths. We call this redshift. Since most galaxies are moving away from us because of the expansion of the Universe, they are all redshifted.

This cosmological redshift tells us about the rate of expansion: the more a galaxy's light is redshifted, the further away it is and the faster it is moving away from us. This behaviour is described in a mathematical relationship (the Hubble–Lemaître Law), which states that the recession velocity of a galaxy, as inferred from its redshift, is equal to its distance multiplied by the Hubble constant, which describes the rate of expansion of space.

KEY DEVELOPMENT

American astronomer Vesto Slipher (1875–1969) first measured the galactic redshifts in 1912, but at the time it was not realized that the 'spiral nebulae' were external galaxies to our own. In the 1920s, Edwin Hubble built on Slipher's work, developing the Hubble–Lemaître Law. Today, astronomers measure the redshifts of galaxies to try to accurately determine the Hubble constant, which describes the rate of the Universe's expansion.

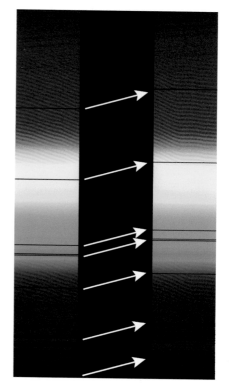

The colours of the spectrum. When light passes through the Sun's atmosphere, some of its wavelengths are absorbed. The left of the diagram shows absorption lines from the Sun, the right shows those for a galaxy that is moving away from us (hence the shift towards the red end).

SUPERNOVAE **p.116**

Stellar Structure

KEY SCIENTISTS: VIRGINIA TRIMBLE • CARL HANSEN • CECILIA PAYNE-GAPOSCHKIN

American astronomer Cecilia Payne-Gaposchkin, whose data were used to determine the paths of stellar evolution. In her postdoctoral thesis she proposed that stars were made primarily of hydrogen and helium.

Photons released by nuclear reactions in a star's core begin a journey to the surface of a star, called the photosphere, which can take up to a million years. It takes so long because of the high density inside the core – photons keep scattering off electrons, so their path to the surface is impeded.

The first zone outside a star's core is the radiative zone, where the star's energy is transported upwards by the photons. The Sun's radiative zone extends for about two-thirds of the Sun's radius. The final third is the convective zone, where the energy is transported by the convection currents that rise up to the photosphere. (Red dwarf stars have no radiative zone – they are entirely convective.)

It's at the photosphere that the energy created by a star is finally released as light, but that light then has to travel through the chromosphere, a narrow, low-density region of gas. Beyond that is the corona, a huge halo of million-degree hot gas that we can't ordinarily see on the Sun, except during a total solar eclipse when the body of the Sun is obscured by the Moon.

KEY DEVELOPMENT

Like all stars, the Sun extends its influence beyond the planets, its solar wind forming a magnetic bubble called the heliosphere. NASA's Voyager 1 and Voyager 2, launched in 1977, passed beyond the heliopause, which is the edge of the heliosphere, in 2012 and 2018 respectively, and they are now in interstellar space.

STARS **p.22** MAIN-SEQUENCE STARS **p.98** RED AND BROWN DWARFS **p.100**

The Mass–Luminosity Relation

KEY SCIENTISTS: JAKOB KARL ERNST HALM • ARTHUR EDDINGTON

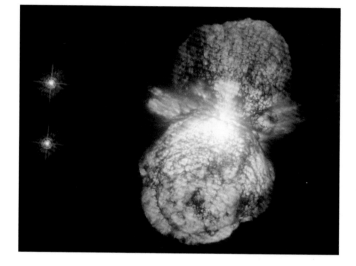

The massive star known as Eta Carinae, which exceeds the mass–luminosity limit.

When we look at the night sky, the stars are all different brightnesses. This is partly because they lie at a variety of distances, but they are also all intrinsically brighter or fainter compared to each other.

There is a pattern to their relative brightnesses. The smallest, least massive stars are the faintest, with the dimmest having less than 1 per cent of the Sun's luminosity, whereas the most massive stars are the brightest, tens of thousands of times more luminous than the Sun. This is described by the Mass–Luminosity relation, first noted by German astronomer Jakob Karl Ernst Halm (1866–1944).

There is a theoretical limit to a star's brightness, known as the Eddington Luminosity after Sir Arthur Eddington (1882–1944), and it describes the maximum brightness a star can reach while maintaining a delicate balance between gravity trying to cause the star to contract, and the pressure of photons created in the core holding the star up. The more massive the star, the stronger the gravity, which means it needs more photons to hold it up. This explains why these stellar behemoths are brighter.

KEY DEVELOPMENT

Not all stars obey the Eddington Luminosity. On occasions, some stars can shine more brightly – especially when undergoing an unstable outburst, as many massive stars are prone to do. An example is the star Eta Carinae which, despite being 7,000 light years away, temporarily became the second brightest star in the sky in the autumn of 1843 when it experienced a violent episode. The cause, to this day, remains uncertain.

STARS **p.22** THE SUN IGNITES **p.50** THE SUN **p.126**

The Power Source of Stars

KEY SCIENTISTS: ARTHUR EDDINGTON • HANS BETHE

KEY DEVELOPMENT

One of the by-products of the proton–proton chain is an unending stream of neutrinos – tiny particles of negligible mass. About 100 trillion of them are passing harmlessly through you every second. Detecting neutrinos from the Sun can tell scientists about the reactions inside the core; neutrinos can also subtly alter geological ores in the Earth, and studying these ores reveals details about the Sun's luminosity in the past.

There are several nuclear reactions that take place inside stars. The Sun's energy is the result of the proton–proton chain. When two protons – that is, two hydrogen nuclei – fuse together, they produce deuterium, also known as 'heavy hydrogen'. Raw energy also spills out and radiates away to be converted into the Sun's light and heat.

The deuterium nucleus fuses with another proton to create a helium isotope called helium-3, plus more energy. Helium-3 is relatively unstable, and can either fuse with another helium-3 nucleus to form helium-4 (stable helium) plus two hydrogen nuclei and more energy, or fuse with helium-4 to form beryllium-7, which then combines with an electron to make lithium. When lithium is added to a proton, it forms two helium nuclei and releases further energy.

Stars more massive than the Sun generate energy through the carbon–nitrogen–oxygen (CNO) cycle, whereby those elements act as catalysts for fusion reactions involving hydrogen, ultimately creating more helium and releasing more energy.

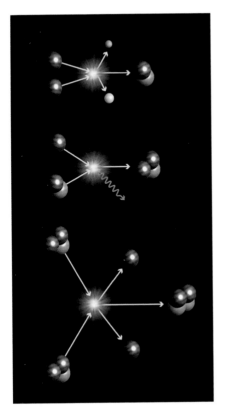

The proton–proton chain, where two hydrogen nuclei fuse to form deuterium (top), then the deuterium fuses with another hydrogen atom to form helium-3 (middle) and finally two helium-3 nuclei fuse to form helium-4 (bottom).

STARS **p.22** THE DEVELOPMENT OF THE SUN **p.55**

Stellar Evolution

KEY SCIENTISTS: JOHANNES KEPLER • FRITZ ZWICKY • WALTER BAADE
RUDOLPH MINKOWSKI

A star's fate depends upon its mass. Our Sun is 4.6 billion years old, and will survive for about another five billion years. The least massive stars, the red dwarfs, can exist for trillions of years, while the most massive have lifetimes spanning just a few million years. Stars with less than eight times the mass of our Sun will run out of hydrogen and evolve into red giants, which eventually blow off their outer layers to form a planetary nebula, leaving behind a white dwarf.

A star more than eight times the mass of the Sun will end its life in a supernova. When it runs out of fuel for nuclear fusion within its core, it experiences catastrophic gravitational collapse. In-falling material inside the star rebounds off the core and sends a violent shockwave propagating outwards that causes the star to explode.

It leaves behind the dense husk of its core, which usually forms a neutron star – an enormously dense object made of neutrons about 10 kilometres (6.2 miles) across. The cores of the most massive stars collapse further under gravity to form black holes.

KEY DEVELOPMENT

On 23 February 1987, a supernova exploded in the Large Magellanic Cloud, which is a nearby dwarf galaxy. Called SN 1987A, the supernova was the closest to us in almost 400 years, and the first nearby supernova of the modern telescopic age. Archive images show the star that exploded; studying those images, and the aftermath of the explosion, has taught astronomers a great deal about how massive stars die.

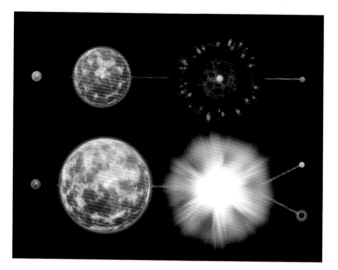

The two evolutionary paths of stars. At top, lower mass stars eventually die as planetary nebulae, while at bottom, massive stars explode as supernovae.

DEATH OF THE SUN **p.57** WOLF–RAYET STARS **p.104** SUPERNOVAE **p.116**

Stellar Nucleosynthesis

KEY SCIENTISTS: FRED HOYLE • MARGARET BURBIDGE • GEOFFREY BURBIDGE
WILLIAM FOWLER

Nuclear fusion is the process by which atomic nuclei are fused together to create a new, more massive atomic nucleus, often belonging to a different element. In 1957, Margaret and Geoffrey Burbidge, William Fowler and Fred Hoyle showed how, through the process of fusion reactions, stars could create many of the elements. They called this process 'stellar nucleosynthesis'.

To generate energy, stars fuse hydrogen into helium. When the hydrogen in their cores runs out, stars then begin fusing helium in a process that results in the creation of carbon.

This is where the reactions in the Sun stop, but in more massive stars they go further. Carbon is fused into oxygen, and so on, resulting in additional layers of silicon, neon and nitrogen until finally the core of the star is comprised of iron, with the other elements around it like layers of an onion.

The energy needed to fuse two atoms of iron together is greater than the energy thereby released, and so the reactions stop. Unable to generate energy, and with energy leaking out of the core via neutrinos, the core of the star collapses. The outer layers follow it, rebound off the condensing core, and explode outwards in a supernova. Temperatures are so great in the supernova – billions of degrees Celsius – that even more massive elements are created.

A supernova explosion is the end result for a star that has fused increasingly heavy elements until it has a core of inert iron.

KEY DEVELOPMENT
Many of the elements formed inside stars are present in the Milky Way Galaxy as dust, but where does this dust mostly come from? Astronomers observing the aftermath of supernova 1987A found the faint infrared emission from enough dust produced by the supernova to build 100 Earth-mass planets. This proves that dying stars are the source of many elements in the Universe.

STARS **p.22** SUPERNOVAE **p.116**

Star Formation

KEY SCIENTISTS: THE HERSCHEL SPACE OBSERVATORY • BART BOK • JAMES JEANS

KEY DEVELOPMENT
It was English physicist James Jeans (1877–1946) who derived the fundamental explanation of how large and massive a molecular cloud has to be to fragment into small star-forming clumps. Only clouds with diameters greater than the Jeans Length (which is dependent on temperature and mass, the required mass being described as the Jeans Mass) will have enough self-gravity to collapse to form stars.

The centre of nearby dwarf galaxy NGC 1569, imaged here by the Hubble Space Telescope, is seen to be a hotbed of star formation.

Stars are born in nebulae – vast clouds of molecular hydrogen gas. You may have seen one yourself – the Orion Nebula, in the constellation of Orion is visible to the naked eye as a fuzzy patch just below the three stars of Orion's Belt. It is 1,344 light years away, a cradle of thousands of hot, stellar infants.

Stars are made when their gas-cloud environment condenses and fragments, usually after it has been disturbed in some way by another passing cloud or the shockwave of a nearby supernova. In order to undergo gravitational collapse, the gas must be cold – just 10 degrees above absolute zero. As it begins to implode, the cloud fragments into individual clumps, each forming a new star or stars.

Eventually, the temperature and density at the core of each clump become so great that the nuclear fusion of hydrogen into helium can occur, a process that releases an abundance of energy. When that happens, a star is born and its heat and radiation eventually blow the remainder of the enshrouding gas away.

THE FIRST STARS **p.43** THE SUN IGNITES **p.50** STAR-FORMING NEBULAE **p.90**

Panspermia

KEY SCIENTISTS: SVANTE ARRHENIUS • FRED HOYLE

KEY DEVELOPMENT
In 1967, NASA launched the Surveyor 3 robotic mission to the Moon. Then, in 1969, Apollo 12 astronauts Charles ('Pete') Conrad (1930–99) and Alan Bean (1932–2018) landed close to where Surveyor 3 had touched down, and retrieved it. On inspection, scientists found that microbes that had initially travelled to the lunar surface on board Surveyor 3 had survived the journey, reanimating when they reached Earth and proving that microbes can survive space travel.

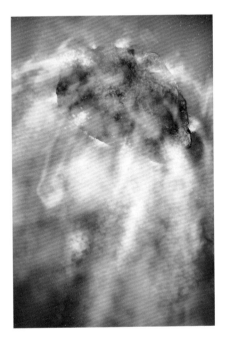

An artist's impression of a comet's nucleus, with dust streaming into space. Is that dust laden with materials needed for life?

Scientists do not yet fully understand how life on Earth began. One controversial theory, called 'Panspermia', is that living things didn't originate on our planet, but were instead brought here by meteorites. There is some circumstantial evidence for this: some of the fundamental building blocks for life, such as amino acids and sugars, as well as complex molecules that facilitate carbon-based chemistry, have been identified on bodies as diverse as meteorites and comets, as well as on the surfaces of icy moons and within the environments of distant star-forming regions.

In theory, microbes would be able to survive the journey through space, perhaps on meteorites blasted off the surface of a planet by the catastrophic impact of an asteroid. Perhaps the inklings of life even travel between the stars on interstellar asteroids and comets, such as 1I/'Oumuamua or 2I/Borisov.

However, despite the suggestion that extreme microbes could survive a journey through space, there is no convincing evidence that we originated from some far-away place in the Universe. Nevertheless, it does seem likely that some of life's building blocks did.

THE MOON **p.138** METEORITES **p.142** COMETS **p.188**

Density Waves and Galaxy Structure

KEY SCIENTISTS: ALAR TOOMRE • C. C. LIN • FRANK SHU

The spiral arms of galaxies are among the Universe's most beautiful and distinctive features. As galaxies rotate, we would expect their spiral arms to wind up, tightening around the luminous galactic core. Why doesn't this happen?

Despite their appearance, spiral arms are not rigid structures that always contain the same stars for billions of years. Instead, the stars are just passing through the arms.

The best way to grasp what occurs in a spiral galaxy is to think of stream of automobiles on a road. Traffic jams occur when vehicles up ahead have to slow down for some reason, causing vehicles behind them to slow in a chain reaction. The automobiles at the front of the jam can then speed up, but meanwhile the jam ripples backwards, slowing and bunching up cars farther behind. A similar thing happens in the spiral arms of galaxies: this phenomenon is known as a 'density wave', which is where the gas, stars and dust tend to bunch up to form the spiral arms.

KEY DEVELOPMENT

Density waves in the spiral disc also have consequences for star formation. As gas bunches up in the arms, their properties reach James Jeans's criteria of mass, diameter and temperature required to spark star formation. As such, the spiral arms of the Milky Way and other galaxies are hotbeds of star-birth, with colourful nebulae glowing with the light of clusters of young stars.

The galaxy Messier 101, also known as the Pinwheel Galaxy, with its prominent spiral arms.

SPIRAL GALAXIES **p.66** THE MILKY WAY **p.84**

Galaxy Evolution

KEY SCIENTISTS: EDWIN HUBBLE • GÉRARD DE VAUCOULEURS

The building of large telescopes in the first half of the twentieth century, such as the 2.5-metre (100-inch) Hooker Telescope on Mount Wilson in California, allowed astronomers to recognize galaxies for what they are – island universes of stars exterior to our own galaxy, the Milky Way. Using their new telescopes, astronomers set about cataloguing all the different types, trying to understand the relationships between the various kinds.

One early effort was the Hubble tuning-fork diagram, which describes different categories of galaxy: spirals, barred spirals, featureless ellipticals and lenticulars, with sub-categories describing the tightness of the spiral arms, or how egg-shaped the elliptical galaxies are. Later, Frenchman Gérard de Vaucouleurs (1918–95) adapted Hubble's diagram to take account of other components of galactic structure, such as ring galaxies and lens-shaped bulges.

Originally, astronomers imagined that elliptical galaxies evolve into spiral galaxies by growing arms. Today, we know it to be the other way around: spiral galaxies form first and later evolve into either dusty lenticular galaxies – a sign that they have run out of star-forming material – or giant elliptical galaxies when they merge with other large galactic structures.

KEY DEVELOPMENT

Galaxy evolution is driven by galaxy mergers, whether they be small galaxies being captured by larger galaxies, or two galaxies of equal mass smashing into each other. Research has found that galaxy mergers were more common in the Universe's past, when galaxies were in general closer together. Because galaxy mergers create star-forming conditions, the peak period of star formation was also around this time, ~10–11 billion years ago.

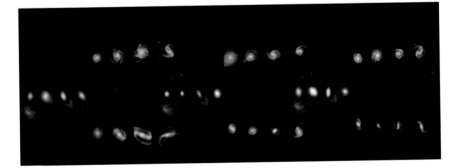

The Hubble tuning fork diagram shows three types of galaxy: ellipticals (left); lenticulars (centre); spirals (right).

SPIRAL GALAXIES **p.66** ELLIPTICAL GALAXIES **p.68** INTERACTING GALAXIES **p.74**

Active Galactic Nuclei

KEY SCIENTISTS: MAARTEN SCHMIDT • CARL SEYFERT • VIKTOR AMBARTSUMIAN
DONALD LYNDEN-BELL

KEY DEVELOPMENT

The sheer power of some AGN was shown in 1962, when Dutch-American astronomer Maarten Schmidt (b.1929) discovered quasars. Their name means 'quasi-stellar object' because they appear as points of light but they are not stars. Closer inspection with larger telescopes showed them to be the active cores of extremely distant galaxies. The discovery of quasars was a crucial step towards understanding the role of supermassive black holes in galaxy evolution.

Galaxies with various AGNs (left to right from top left): bright core with dust ring; dust ring and accretion disc around a black hole; a brilliant star core; a bright star core; side-on view of galaxy showing emittent beams of ionized matter; the same galaxy viewed from Earth.

Scientists predicted that a giant black hole lies at the heart of every large galaxy long before it was ever proven to be the case.

In the 1940s, astronomer Carl Seyfert (1911–60) noticed that some galaxies had unusual emissions from their cores, as though something incredibly bright lurked there, occasionally partially hidden by cosmic dust. In the 1950s, surveys of the Universe identified distant objects radiating powerful radio waves, and in the 1960s quasars, which are optical counterparts of these radio sources, were identified.

Astronomers began piecing together the evidence to build a theory of active galactic nuclei (AGN), according to which at the centre of each galaxy is a supermassive black hole, and some of these holes are gobbling

matter. Black holes draw in gas, which forms a spiralling 'accretion disc' with temperatures of billions of degrees Celsius. It is these discs that we see radiating in the centre of AGN. Quasars are the brightest of them.

The energy they release is believed to have helped end the cosmic dark ages by ionizing hydrogen gas in space. In the most extreme cases, the magnetic fields belonging to AGN can whip up some of the gas and funnel it away in jets of charged particles moving at nearly the speed of light. When the jet is pointed directly at us, the AGN appears incredibly bright, and we call such objects 'blazars'.

RADIO GALAXIES **p.78** QUASARS AND BLAZARS **p.81**

Dark Matter

KEY SCIENTISTS: VERA RUBIN • FRITZ ZWICKY

Some 85 per cent of all the matter in the Universe is invisible to us. We call it dark matter, and we only know that it might be there because of its gravitational influence.

Think of a disc galaxy, made up of stars orbiting its centre. You would naturally expect the stars at the outermost edge to complete a lap much more slowly than the ones closer in. Scientific observations, however, show that stars orbit at about the same speeds, regardless of their distance from the centre. The explanation seems to be that there is dark matter present, and it is that which provides additional gravity.

Unfortunately, no one currently knows what dark matter actually is. Studies at CERN's particle-smashing Large Hadron Collider have consistently drawn a blank. Our best guess is that it's built out of supersymmetric particles, opposite to but more massive than those that make up the Standard Model.

KEY DEVELOPMENT

Not all scientists believe that dark matter is real – a minority instead believe that a tweak to Newton's laws of gravity at low accelerations could mimic the gravitational attraction of dark matter. They call this theory Modified Newtonian Dynamics (MOND). While it has had some successes, for example in explaining the accelerations of stars as they orbit galaxies, the consensus view is still that dark matter is real.

A false colour, multi-wave-length image of a merger between two galaxy clusters. The blue glow represents the implied location of dark matter in the clusters.

CLUSTERS AND SUPERCLUSTERS **p.64** SPIRAL GALAXIES **p.66**
THE HALO AND GLOBULAR STAR CLUSTERS **p.86**

Dark Energy

KEY SCIENTISTS: ADAM RIESS • SAUL PERLMUTTER • BRIAN SCHMIDT

Dark energy is the force that's accelerating the expansion of the Universe. It makes up a whopping 68.3 per cent of all the mass and energy in the cosmos, but despite its huge influence on space and time, it's still a largely unknown entity.

Its existence was discovered by chance in 1998. While astronomers including Nobel Prize-winning Brian Schmidt (b.1967), Saul Perlmutter (b.1959) and Adam Riess (b.1969) were using the light of Type Ia supernovae (exploding white dwarf stars in distant galaxies) to work out how fast their galaxies are moving away from us, they noticed something strange: some of the supernovae were dimmer than they should have been, meaning they must be farther away than expected. The Universe wasn't just expanding – its expansion was accelerating. Given the name 'dark energy', the force behind this acceleration started to take over between seven and eight billion years after the Big Bang. If dark energy continues to accelerate space's expansion unabated, it will decide the fate of the Universe, in a 'Big Rip'.

KEY DEVELOPMENT

Albert Einstein developed general relativity before it was known that the Universe is expanding, so when his equations showed that the Universe must be expanding, he assumed them to be wrong and introduced a 'cosmological constant' to counteract it. When he realized that the Universe does expand, he removed the cosmological constant, but it turns out that dark energy could be the cosmological constant in another guise.

A speculative image of the visible cosmos with our galaxy at the centre.

THE BIG BANG **p.35** THE MOMENT OF INFLATION **p.36**

Index

Picture Credits